Hispanic Ecocriticism

STUDIES IN LITERATURE, CULTURE, AND THE ENVIRONMENT

Edited by Hannes Bergthaller, Gabriele Dürbeck,
Robert Emmett, Serenella Iovino, Ulrike Plath

Editorial Board:
Stefania Barca (University of Coimbra, Portugal)
Axel Goodbody (University of Bath, UK)
Isabel Hoving (Leiden University, The Netherlands)
Dolly Jørgensen (Umeå University, Sweden)
Timo Maran (University of Tartu, Estonia)
Serpil Oppermann (Hacettepe University, Ankara, Turkey)
Dana Phillips (Towson University, Baltimore, USA)
Stephanie Posthumus (McGill University, Montreal, Canada)
Christiane Solte-Gresser (Saarland University, Saarbrücken, Germany)
Keijiro Suga (Meiji University, Tokyo, Japan)
Pasquale Verdicchio (University of California, San Diego, USA)
Berbeli Wanning (University of Siegen, Germany)
Sabine Wilke (University of Washington, Seattle, USA)
Hubert Zapf (University of Augsburg, Germany)
Evi Zemanek (University of Freiburg, Germany)

VOLUME 6

José Manuel Marrero Henríquez (ed.)

Hispanic Ecocriticism

Bibliographic Information published by the Deutsche Nationalbibliothek
The Deutsche Nationalbibliothek lists this publication in the Deutsche Nationalbibliografie; detailed bibliographic data is available in the internet at http://dnb.d-nb.de.

Library of Congress Cataloging-in-Publication Data
A CIP catalog record for this book has been applied for at the Library of Congress.

Cover illustration:
Jesús de la Rosa: Telecommunications Dance (2006).
Waxes on wood, 60 x 42 cm.
Courtesy of the artist.

ISSN 2365-645X
ISBN 978-3-631-78550-8 (Print)
E-ISBN 978-3-631-78955-1 (E-PDF)
E-ISBN 978-3-631-78956-8 (EPUB)
E-ISBN 978-3-631-78957-5 (MOBI)
DOI 10.3726/b16211

© Peter Lang GmbH
Internationaler Verlag der Wissenschaften
Berlin 2019
All rights reserved.

Peter Lang – Berlin · Bern · Bruxelles · New York · Oxford · Warszawa · Wien

All parts of this publication are protected by copyright. Any utilisation outside the strict limits of the copyright law, without the permission of the publisher, is forbidden and liable to prosecution. This applies in particular to reproductions, translations, microfilming, and storage and processing in electronic retrieval systems.

This publication has been peer reviewed.

www.peterlang.com

Acknowledgments

I would like to thank Hannes Bergthaller for his encouragement to pursue a book that, limited to a few topics, aspires to reflect the vast array of issues that are key to the development of ecocriticism in Hispanism and in literary and cultural studies in general. I am also grateful to Gabriele Dürbeck, Robert S. Emmett, Serenella Iovino and Ulrike Plath for their accomplished academic work in the field of environmental humanities. Their initiative is much needed in the Anthropocene era, and both culture and nature benefit from the series *Studies in Literature, Culture, and the Environment* that they have created. I must also thank every person involved in this project: Natalia Álvarez Méndez, Arturo Arias, Laura Barbas-Rhoden, Scott DeVries, Gisela Heffes, Jorge Marcone, Pamela Phillips, Beatriz Rivera-Barnes, and Manuel Silva-Ferrer, who kindly allowed themselves to be distracted from their ongoing projects and responded to my call for a contribution to *Hispanic Ecocriticism* with their work. I thank Jesús de la Rosa for his artwork for the front cover of this book. Ellen Skowronski-Polito has been very generous with her time and a devoted and patient assistant editor who once again has accepted with serene temperance every single question I have had to make about English, a language familiar to me but nonetheless not my own. And thanks to Luis and Marta Marrero, Susanne Esser, Pilar Henríquez Ponce and the memory of José Luis Marrero Cerpa for being the reason to persist in every endeavor of life.

Contents

José Manuel Marrero Henríquez
Introduction: On Hispanic Ecocriticism 9

Hispanic Ecocritical Theory

José Manuel Marrero Henríquez
Ecocriticism of the Anthropocene and the Poetics of Breathing 19

Jorge Marcone
Towards an Amazonian Environmental Humanities 49

Laura Barbas-Rhoden
Gendering Ecohispanisms: Knowledge, Gender, and Place in a
Pluricultural Latin America 69

Spanish Ecocriticism

Pamela Phillips
Enlightening Nature: An Ecocritical Reading of Eighteenth-Century
Spanish Literature 93

Natalia Álvarez Méndez
Subject and Landscape: Encounters with Nature in Contemporary
Spanish Narrative 113

Latin American Ecocriticism

Scott DeVries
The Quiroga Frame: Animal Studies and Spanish American Literature 137

Arturo Arias
Indigenous Knowledges and Ecological Thought: Jak'alteko Maya
Victor Montejo's Fables 157

Beatriz Rivera-Barnes
Sadder Tropics: The Hate of Nature in Juan José Saer's *El entenado* and
Dorian Fernández-Moris's film *Desaparecer* 179

Gisela Heffes
Exclusive Natures: Latin American Cities in Urban Ecocritical
Perspectives 199

Manuel Silva-Ferrer
Petrofictions: Nature and Imaginaries of Oil in Latin America 225

Contributors 247

José Manuel Marrero Henríquez

Introduction: On *Hispanic Ecocriticism*

In his text "The Analytic Language of John Wilkins", Jorge Luis Borges highlights how the great virtues of natural languages are to be found in their defects. Within the story's title, the reference to English natural philosopher and writer John Wilkins recalls the seventeenth century innovator's attempt to remedy linguistic defects by creating a perfect language, one that would provide a sign for everything in the world, from the most abstract to the most miniscule. Not allowing for any ambiguity, Wilkin's proposed language consisted only of precise words inserted in a grammar as accurate as the metric system in which words behave like numbers with exact meaning and self-containing definitions; however, words that act like numbers and perfect signs for perfect meanings ultimately result in nonsense. This ideal but impossible language suits one of Borges's characters, Funes, the boy who, after an accident, develops a boundless memory capable of having a term for every occurrence: the name for the fly that is here and a different name for the same fly over there, or an expression for the leaf in the air and another for same one once back on the earth. Perfect as it appears to be, Wilkins's language needs the common approval of his departing classification of the world, and only a memory as huge and worthless as Funes's is able to learn it. Borges uses Wilkins's project to praise the limitations of human beings and the imperfections of natural languages, despite their being full of inexactitudes and the illogical categories of grammatical genres, ambiguities, homonyms, synonyms, and so forth. Those same imperfections, which are able to produce an infinite quantity of messages with a limited number of signs and sounds, are also the pillars of art and literary creation.

Since its inception, *Hispanic Ecocriticism* has aimed to reproduce this basic condition of natural languages by making its imperfections also its virtues. On one hand, the title *Hispanic Ecocriticism* is too broad a designation to be able to be true, for there is no possible way to condense into one book all the literary variations and critical viewpoints that are important from an ecocritical perspective within Hispanism. On the other hand, the adjective "Hispanic" is overly restrictive inasmuch as it obscures the fact that the ecocritical topics, theoretical reflections, and critical proposals within Hispanism are also of general interest to ecocriticism as a global and transnational movement. Furthermore, the term

"Hispanic" treats the adjective as if it were a pristine and crystal clear term without internal conflicts. The adjective "Hispanic" tends to avoid the reflection of its own meaning, which, without any doubt, points to the conflict of the metropolitan powers of Spain—and by extension, Portugal, England, France, the Netherlands, and other European nations—and their part in the imposition of the Spanish language and political powers over the native languages and cultures of the peoples of America.

Nonetheless, just as Borges affirms the limits of natural languages, these contradictions and excesses in scope, which are simultaneously reductionisms, tend to erase the historical footprints of cultural conflict in favor of harmonious images. In brief, the insufficiencies of the title are also its virtues, for not only does it allow *Hispanic Ecocriticism* to cover a vast array of issues that are relevant in ecocriticism as a general movement and as a Hispanic field of research, but it simultaneously facilitates recurring dialogue of complementarity and collaboration, thereby giving space to justice and encouraging Hispanism's effort to undress its arrogant apparels and humbly accept that its language and culture can no longer be considered without the influence of indigenism, its counterpart in Latin American culture and political history.

In the diary of his first journey, Columbus, inspired by the Americas' magnificent landscapes and the beauty of the inhabitants, wrote that he had discovered Paradise. By the second journey, however, a different imagination prevailed, one that envisioned this Garden of Eden as a profitable land of immense natural and human resources that could be acquired through the alphabetization of its people. In the years that followed, Spain capitalized on this vision, conquering and colonizing for its own benefit and for the advantage of other European countries. The transatlantic inception of Columbus's travel diary placed the Spanish language in a complex field of conflicting interests upon which it has built its modern idiosyncrasy, encompassing multifaceted American and European identities, contending realities, and textual *loci*.

The Spanish language has been adapted throughout the centuries by the inhabitants and cultures of a multitude of American sites, from regions of sublime romantic wonder akin to Western imagination to distressful jungles and formidable rainforests, from cosmopolitan cities with European flavor to suburban populations of indigenous alluvial peasants, from rural lands of creole influence to Caribbean islands of African ancestry, the Andes, the Amazonia, and the remote ecosystems of Patagonia. In a restorative circle that has yet to be completed, the Spanish language places its alphabet and phonetic system at the service of the Mapuche and other American ancestral cultures to ease the written survival of their oral traditions.

The Spanish language has enriched its European roots with American branches and fruits that are becoming prominent in what should be now a new design of the Hispanic literary tradition. Languages have a great capacity to evolve, adapt, clash, and intermingle along with the circumstances of their history, and it is in America where Spanish, imbued with its influence of European culture and the ancestral inhabitants of America, for the first time gives birth to texts able to cope with the unprecedented circumstances of the Anthropocene. First developed in Latin America and later in Spain, ecocriticism has found in this complex, ecumenical, and democratic Hispanism a rich field of research that transcends frontiers to acquire a dimension of global interest.

In a renewed exchange of transatlantic relationships, the primary ecocritical impulse within Hispanism comes from Latin America, for Spanish academia has demonstrated reticence to incorporate the environmental agenda in the Humanities. In Latin America, topics of great relevance include the oil industry and environmental contamination, the use and abuse of forests and rivers, gender and the perception of environment, urban ecology, indigenous thought about the land and its human and nonhuman inhabitants, the African factor in the perception of the environment, suburban ecosystems, and waste and social justice. In Spain, ecocriticism turns its gaze to animal rights and the ecological footprints of human activity in contemporary narratives of eco-science fiction, in dystopias, and in literature inspired by natural or rural landscapes that conceal ways of life and cultures in peril of extinction.

Old *loci* appear under a new light, and new *loci* arise. Hispanic ecocriticism has a vast agenda in building a tradition that reconnects Latin America to Spain, ancestral and contemporary knowledge, past periods of literature with new literary trends, and contemporary ecological sensibility with animal ethics. When collaboration challenges competition, technology can serve traditional agriculture and environmental health, European degrowth and slow food movements see an ally in the American *sumak kawsay*, and the indigenous perspectives on the American land enrich contemporary animal and landscape ethics. With a collaborative attitude, Hispanic ecocriticism finds a rich soil in the main topics of environmental concern in the literature of Latin America and Spain, not only as a source for renewing critical analysis and hermeneutics, but also for the benefit of global environmental awareness.

Under the same strength of will with which natural languages aim to give an accurate account of the reality surrounding their utterances, even if that aim is not possible to be achieved in its entirety, this manifold agenda is carried out with an inclusive, just, and democratic perspective on Hispanism. Three chapters of *Hispanic Ecocriticism* focus on literary theory. José Manuel Marrero Henríquez develops a "poetics of breathing" as a general ecocritical theory able to sustain a coherent ecocritical practice in a variety of cultural traditions. Whether the texts under scrutiny be American or European, part of the oral tradition or fully inserted in the literate tradition, akin to indigenous cosmogonies or Western mindsets, Marrero Henríquez's "poetics of breathing" inspires ecocriticism to explore the transbordering possibility of how words breathe or, namely, the process by which literature proves to be the ultimate result of the natural evolution that rewards those who are able to grasp its beauty and regularities in time and space. Despite being aerial, breathing is a solid base from which to discover concomitance and find the common roots of a vast array of cultural expressions inside—and outside—Hispanism.

In his theoretical proposal on the Amazonia, Jorge Marcone considers the systematic absence of Amazonian texts in the canon of every Andean nation as a symptom of the insufficiencies of the academic acceptance of transcultural theories in countries characterized by the coexistence of a plurality of ontological worlds. As a departure point for his study, Marcone refers to the fact that the Amazon and its denizens have been considered by the Regionalist novel as a site for exploitation and the "Other" in contrast to the European and *criollo*. From this perspective, Marcone studies how Amazonian ontologies have contributed to the ecocritical debates on the status of nonhuman beings and how these ontologies promote a challenging revision of the linguistic consideration of these beings whose very enunciation could be thought of as more-than-human. The very notion of text in literary theory ventures beyond the split between nature and society and human and nonhuman, and the foundation of discursive elaborations on literary history is questioned.

Like the notion of text, that of gender becomes a powerful dynamic category, especially when it intersects with humanity, ethnicity, race, class, age, and geopolitical location. Laura Barbas-Rhoden studies gender as a theoretical means to prevent the pluricultural context of Latin America from being blurred or obscured in journalistic and academic discourses. Barbas-Rhoden centers her attention on the interpretation of Latin American cultural texts and makes of gender and ecocriticism tools that illuminate every cultural artifact, not only those signed by a gendered qualification, as it happens, for example, with explicitly feminist production, but also key topics of Latin American culture such as that of civilization and barbarism.

Oppositions are a powerful hermeneutical tool and a habit of the intellectual mind. They have existed and exist, but they may be reread under new lights, lights that are able to raise doubts on the clear and well-defined drawing of their borders. This effect happens not only in ecocritical theory, when Marrero

Introduction: On *Hispanic Ecocriticism* 13

Henríquez introduces a binding poetics of breathing, when Marcone reevaluates Amazonian ontologies or when Barbas-Rhoden multiplies Latin American gender cultural overtones, but also occurs in ecocritical practice, such as when Pamela Phillips studies Enlightenment and examines its influence on Spanish literature. In historical terms, literature of the Enlightenment has been marginalized as a second-class period in which nature appears as an object progressively controlled by humankind. Nonetheless, Pamela Phillips not only studies the aesthetic value of nature in Spanish literature of the Enlightenment, but also sees in this period the beginning of the acknowledgement of nature as a limited resource. Phillips proposes Enlightenment as a valuable literary period for ecocritical consideration and dialogue with contemporary Spanish literature of the twenty-first century both within Hispanism and as a general ecocritical issue. Through her reading of Jovellanos, Feijoo, and eighteenth century Spanish poetry, Pamela Phillips studies the ecological concerns of Spanish Enlightenment that are of great relevance to contemporary ecocritical thought.

It is precisely in the respect for nature and return to rural life that Natalia Álvarez Méndez finds a common ground for a significant number of Spanish novels from recent decades. Rurality appears in the works of Llamazares, Mateo Díez, Merino, Jenn Díaz, Manuel Darriba, Lara Moreno, Iván Repila, and other authors as a challenge to the failure of modernity and its economic and moral crisis. Rural life develops apart from consumerism and allows for the depiction of agricultural landscapes, the description of life and work in small villages, and the denouncement of the ecological destruction of nature. Natalia Álvarez Méndez's landscapes are, like animals, biological beings threatened by capitalism, consumerism, and destructive industrialization and can serve as a bridge to Phillips's study of eighteenth century Spanish poets.

On Latin American shores and under the spotlight of contemporary debates on animal ethics, Scott DeVries centers his work on the study of animal representation as a way to reconsider the historical system in which Latin American literature is studied and classified. DeVries explores the representation of animals in fiction and poetry from the major Latin American literary periods, the nineteenth century to contemporary narrative, from Faustino Sarmiento to Luis R. Sepúlveda and Fernando Raga, in light of animal ethics and the concerns of present debates on animal studies. DeVries's chapter demonstrates how his "fauna-criticism" is not only a study of animal representation in Latin American literature, but also allows a reframing of Latin American literary history according to the animal-centered ethical position of poetry and fictional prose.

Similarly, identity shared by human and nonhuman animals is a key topic in Arturo Arias's study of ecological thought in indigenous knowledge through his reading of the fables of the Jakaltek Maya author, Víctor Montejo. Arias finds in Montejo's fables an integrated eco-space of animal subjects, human and nonhuman; indeed, their natural and supernatural environment, in a very localized emplacement, demonstrates the dynamics of life in a web formed by relationships among humans, animals, plants, natural forces, spirits, and landscapes. If DeVries's "fauna criticism" questions human identity and Arias's reading of Víctor Montejo tends to erase the divide between nature and culture by envisioning cultural products as natural processes, Beatriz Rivera-Barnes's reading of Juan José Saer's *El entenado* and Dorian Fernández-Morris's film *Desaparecer* take the opposite direction by exploring the radical distance that flourishes between humans and nature when the hate of nature is the central feeling of an artwork.

An undesired feeling, hate hides the human need to understand nature and reconnect with the Earth, its geographical features, its minerals, vegetables, animals, and every one of its inhabitants. This need not only affects the reconfiguration of a Hispanic literary history, the creation of new texts, and the renewal of the readings of its canonical texts, but it also stimulates the envisioning of possibilities of social life that respond to the ecological crisis that defines the era referred to as the Anthropocene.

Responding to aggression against ecological values, the final two chapters reflect on the literary and artistic expressions that confront environmental problems in urban settings and the global threats of industrial activities in Latin America. Gisela Heffes studies how the city, as an immense biological organism, copes with the environmental perils of industrial development by turning knowledge into a tool to convert Latin American cities into green spaces where the global problems of wealth, segregation, and waste management can be solved. Manuel Silva-Ferrer studies literature inspired by the rise and expansion of fossil energies through the lens of the ecological turn. Reviewing a number of twentieth century Latin American texts in which the petroleum industries play a relevant role, Silva-Ferrer shows the contradictions between local struggles and global flows, thereby relocating petroleum literature under the light of environmental global awareness and an ecocritical perspective.

Although Borges recognizes the imperfections of natural languages, he also sees in them a great tool for human communication and artistic creation: a limited number of elements for an infinite array of expressions. The way to solve *le défaut des langues*, against which Mallarmé warned, is not through John Wilkins's and others' path to create universal languages following mathematical patterns and perfect definitions. Rather, the answer is to delve deeper into those defects, working on them and turning them into marvelous virtues.

This response reflects the utopia behind any language as well as the objective of *Hispanic Ecocriticism*, which aims to provide access to a wide range of cultural topics in Hispanic ecocriticism through a limited number of specific issues. Regarding language, Borges quotes Chesterton to comment on how these issues are "more bewildering, more numberless, and more nameless than the colors of an autumn forest [...] can every one of them, in all their tones and semitones, in all their blends and unions, be accurately represented by an arbitrary system of grunts and squeals. He believes that an ordinary civilized stockbroker can really produce out of his own inside noises that denote all the mysteries of memory and all the agonies of desire". Like the scarce letters of the alphabet that are the basis for language, poetry, and the monument of culture, the finite chapters of *Hispanic Ecocriticism* on specific and selected topics aspire to give a general account of the multiplicity of issues that affect not only Hispanism, but also ecocriticism as a general, global trend aiming to understand the literatures and cultures of the Anthropocene and the societies in transition to a post-carbon civilization.

Hispanic Ecocritical Theory

José Manuel Marrero Henríquez

Ecocriticism of the Anthropocene and the Poetics of Breathing[*]

Abstract: If culture is part of nature, the hypothesis that written forms of Hispanic literary texts preserve the desire to grasp the immediate knowledge of the environment, which Walter Ong studies in primary orality or Bruno Latour describes in the paradoxical movement of scientific investigation by which one moves away from reality to approach it, points to a commendable theoretical and critical objective. In Hispanic literatures, Latin American or European, those akin to indigenous cosmogonies, Christian lineage or declaredly atheist, or of oral, written or digitally virtual origin, Anthropocene ecocriticism has the important generic and transboundary task of highlighting the complementarity of artistically complex literary and oral processes and the poetics of breathing that they share.

Keywords: Ecocritical Theory, Poetics, Breathing, Hispanism, Latin American Literature

Introduction

The objective of the following pages is to design a poetics of breathing that serves as a general ecocritical literary theory capable of responding to the global challenges of the Anthropocene era within the scope of Humanities. Wherever problematic aesthetic and ideological differences arise, such poetics will be inclusive and comprehensive as it seeks to find common ground by investigating potential complementarities. The poetics of breathing aspires to provide the resolution of conflicts, envisions the blurring of the definite and clear limits that literary criticisms based on irreconcilable binary oppositions draw, and, in the formation of its design, appeals both to works of Latin American literature, which thrive on the diverse imaginaries and traditions of the original peoples of America and the ideologies or worldviews they support, and to European and Latin American literary works based on Western-rooted imaginaries and

---

[*] This chapter is part of the project "Environmental Humanities. Strategies for Ecological Empathy and the Transition towards Sustainable Societies" (HUAMECO), subproject 2, "Stories for Change": HAR2015-67472-C2-2-R (MINECO / FEDER).

traditions. Both traditions and the various hybridizations that have resulted shape the field of Hispanic literature in Europe and America, and all are relevant if one wants to design a poetics of breathing that serves as a general theory of ecocriticism.

The poetics of breathing does not hesitate to support itself through literary examples that can be considered opposites, and it responds positively to each of the rhetorical questions detailed below. On both sides of the Atlantic, does not a naturally common breath pulsate? Is not literate culture, even the most urban, digital, and technologically advanced, a sophisticated manifestation of evolved nature? Does not the continuous rhythm of inspiration and exhalation from which both oral and written poetry are nourished everywhere in all of nature and its spatial regularities (meanders, geological strata, fractal structures) and temporal regularities (tides, seasons, fertility cycles) originate from an Earth that also breathes? Is not capturing that rhythm the privileged object of inquiry for both ancestral peoples of oral culture and the highly technological and literate?

The poetics of breathing assumes that orality persists in writing in the same way that the global visions provided by satellites corroborate the primitive intuition that everything is interrelated and that human beings do not govern but rather form part of Earth's ecosystem in its entirety. Literary inquiry, even the most technical, formal, and unconcerned about matters relating to nature and its biological, historical and political circumstances, preserves to some extent its anchor in nature, which is fundamental in oral literatures.

The poetics of breathing is built on nature, without fear of error and with certainty that nature is a space that literary and oral traditions, Amerindian and European, share. It is not about erasing cultural and literary differences, camouflaging conflicts, nor obviating the distance between the executioner to the victim; rather, it is about promoting cooperation on common ground and looking for ecumenical responses in an era like the Anthropocene in which there are no exclusively independent matters because local facts are also global and concern humanity as a whole. Inspired by the rhythmic and harmonic regularities of nature, the poetics of breathing, with declared willingness, reads literary works as belonging to the great homeostatic ecosystem of culture and considers that an Anthropocene ecocriticism has the relevant task, generic and transborder, to highlight the complementarity of the artistic and oral art processes, whether they be within Hispanic, Latin American or European literatures, related to indigenous cosmogonies, Christian lineage, or declared atheism, or of oral, written, or digitally virtual roots.

Indigenism and Ecology

To illustrate the differences that unfold in the ecocriticisms of Latin American hispanism, it is worth mentioning Paredes and McLean's typology of true ecologist Latin American literature, which founds itself on a fundamental dichotomy of American literary imaginaries. This dichotomy counterposes the indigenous tradition to the Hispanist (and Europeanist) and allows Paredes and McLean to discriminate Spanish literature and part of canonical Latin American literature from the Hispanic ecologist literature that its typology proposes.[1]

Paredes and McLean's reasoning is definite and unwavering. The Judeo-Christian imprint has fostered a negative attitude towards nature and has privileged the role of the human being in the destinies of the universe. The European imprint, identified by Paredes and McLean through humanism, rationalism, capitalism, productivist socialism, and positivism, has formed a skein of cultural attitudes that intertwines machismo, racial discrimination, ecological insensitivity, cultural arrogance, and the universalist discourses of progress and scientism (p. 25).

Both imprints have permeated Latin American literature and emerge in the American literary traditions that Paredes and McLean consider alien to ecological literature. With few exceptions, neither the "novels of the earth" nor the novels of "magical realism" can be considered as belonging to ecologist literature. In the novels of magical realism, the European gaze hallucinates over an American space that it cannot fully understand. In the novels of the earth, the dichotomy of "civilization / barbarism" that comes from the Argentine romantics weighs heavily and frames Nature in a dialectic in which the human being has the moral right to "transformar y destruir la fuente del subdesarrollo,

---

[1] Paredes, Jorge / McLean, Benjamin: "Hacia una tipología de la literatura ecologista en el mundo hispano". *Ixquic* 2, 2000, pp. 1–37. On the usefulness and inadequacies of the Paredes and McLean typology, see Marrero Henríquez, José Manuel: "Ecocrítica e hispanismo". In: Flys Junquera Carmen et al. (eds.): *Ecocríticas. Literatura y medio ambiente*. Iberoamericana Vervuert: Frankfurt 2010, pp. 193–218; and Marrero Henríquez, José Manuel: "Pertinencia de la ecocrítica". *Revista de Crítica Literaria Latinoamericana* 79, 2014, pp. 57–78. In a line that indicates the convenience of a transnational and inclusive ecocriticism, see Heffes, Gisela: "Introducción. Para una ecocrítica latinoamericana: entre la postulación de un ecocentrismo crítico y la crítica a un antropocentrismo hegemónico". *Revista de Crítica Literaria Latinoamericana* 79, 2014, pp. 11–34; and Barbas-Rhoden, Laura: "Hacia una ecocrítica transnacional: aportes de la filosofía y crítica cultural latinoamericana a la práctica ecocrítica". *Revista de Crítica Literaria Latinoamericana* 79, 2014, pp. 79–98.

el primitivismo y la maldad, que toma forma de bosques, animales, ríos, selvas, montañas [y] lagos" (p. 21) ["transform and destroy the source of underdevelopment, primitivism and evil, which takes the form of forests, animals, rivers, jungles, mountains [and] lakes"].[2]

In the ecological imaginary of this delineation, the possibility of an ecological literature is reduced to the scope of indigenous cultures of Mesoamerica because these cultures clearly defy the historical, political, religious and philosophical metanarratives imposed by European imperialism after 1492. Given the contemporary boom of literary studies of poets from the original cultures of America, those poets who work within the original language could be added to Paredes and McLean's list of ecologist authors; these include Briceida Cuevas Cob, Maya-Yucateco; Natalio Hernández, Nahuatl, also from Mexico; Odi González, Quechua of Peru; Leonel Lienlaf, Mapuche Mapudungun. Other poets who write in Spanish, but from indigenous traditions, could also join: Rosa Chávez (Mayan, Guatemala), Jaime Huenún (Mapuche-Williche), Roxana Miranda Rupailaf (Mapuche-Williche), and Liliana Ancalao (Mapuche, Argentina).[3]

Hispanism and Ecology

Useful for its precision, Paredes and McLean's typology privileges the ecological potential of literatures rooted in indigenous cultures, but does not account for the possibilities of an environmental literature and criticism in texts of European affiliation, Spanish or Latin American. There is no doubt that a Spanish ecologist literature is possible, just as it is possible to extract foundations for an outline for a tradition of environmental assertion from the Judeo-Christian tradition, the progressive European scientific tradition, or from the Latin American ecologist literature within the European tradition. Similarly, it is possible to find readings of an ecological nature in Hispanic literature with Judeo-Christian and European roots.

---

2 Except in those cases in which specific translations are referred to in notes, all translations from Spanish into English belong to Ellen Skowronski-Polito and José Manuel Marrero Henríquez.
3 See Pierce, J. M.: "Entrevista a Arturo Arias y Luis Cárcamo Huechante". *Pterodáctilo* 9, 2010, pp. 2–8, from http://pterodactilo.com/numero9/files/2010/11/Pterodactilo_09_DossierEspecial.pdf (23-11-2015). Joseph M. Pierce interviews Arturo Arias and Luis Cárcamo Huechante for a discussion on the different poetics of contemporary authors based on the cultures of the native peoples of America and the problems involved in choosing a language as a vehicle for writing.

In fact, it is precisely European scientific development that gives rise to the term "ecology", which, starting with botanical studies that initially focused on mere taxonomy, benefits from the understanding that plant physiognomy depends on factors such as soil or climate. In this physiognomy, beings from several realms come into play, and increasingly complex realities drive to make "ecosystem" a unit of study and, ultimately, the entire planet as a huge ecosystem. Similarly, scientific development has allowed the contemplation of planet Earth as an immense ecosystem in which all events are interrelated, which includes: the first moon landing (1969), the creation of the Internet (1969), the first global study of the world production and consumption system, *The Limits of Growth* (1972), the foundation of the first environmentalist party (1972), the creation of the World Wide Web (1989), the Kyoto Treaty, the new indices to measure the progress of nations, and the scientific perspectives formed by the United Nations' Intergovernmental Panel on Climate Change. Without contradicting or undermining these facts, an environmental literature inspired by indigenous world visions and an ecocriticism dedicated to it are possible.[4]

There are noteworthy examples of ecological claims encouraged by the fear of the success of progress from within the culture of progress. Neruda, disenchanted with capitalist development and the deification of communist leaders, translates his socio-political alignment to a clearly ecological orientation. This transformation happens in poems such as "Se llenó el mundo" ["The World Filled Up"] from *Fin de mundo* [*World's End*] (1969), in which the critique of progress comes from whomever belongs to the culture of progress and its imaginary:

> Hermosos fueron los objetos
> que acumuló el hombre tardío,
> el voraz manufacturante:
> conocí un planeta desnudo
> que poco a poco se llenó
> con los lingotes triturados,
> con los limones de aluminio,
> con los intestinos eléctricos
> que sacudían a las máquinas
> mientras el Niágara sintético
> caía sobre las cocinas.

---

4   On the European context of scientific development in relation to the advancement of a global ecological consciousness, see Marrero Henríquez, José Manuel (ed.): *Literatura y sostenibilidad en la era del antropoceno*. Fundación MAPFRE-Guanarteme: Las Palmas de Gran Canaria 2011a.

Ya no se podía pasar
en mil novecientos setenta
por las calles y por los campos:
las locomotoras raídas,
las penosas motocicletas,
los fracasados automóviles,
las barrigas de los aviones
invadieron el fin del mundo:
no nos dejaban transitar,
no nos dejaban florecer,
llenaban arenas y valles,
sofocaban los campanarios:
no se podía ver la luna. (p. 98)

[Beautiful were the objects
the late man accumulated
the voracious manufacturer:
I knew a naked planet
that little by little filled
with crushed ingots,
with aluminum lemons,
with electric intestines
that shook the machines
while synthetic Niagara
cascaded in the kitchens.

No one was able to pass
in nineteen seventy
through the streets or beyond the fields:
the threadbare locomotives,
the distressed motorcycles,
the failed automobiles,
the bellies of the planes
invaded the end of the world:
they did not allow us to travel,
they did not allow us to flower,
they filled the sands and the valleys,
they suffocated the bell towers:
no one could see the moon]. (p. 159) [5]

---

5   Neruda, Pablo: *Fin de mundo*. Losada: Buenos Aires 1969; Neruda, Pablo: O'Daly, William (trans.): World's End. Copper Canyon Press: Washington 2009.

This poem by Neruda criticizes the variegation of an overproduction that leaves the human being and nature a tiny healthy space in which travel is difficult. Clearly, the poem has an environmental context outside of an indigenous worldview, because it is consistent with a critical understanding of the progress and social advancement of those who live immersed in the culture of development. Similarly, this view can be found in Huidobro's *Altazor* (1931). His verses read:

> Después de mi muerte un día
> El mundo será pequeño a las gentes
> Plantarán continentes sobre los mares
> Se harán islas en el cielo
> Habrá un gran puente de metal en torno a la Tierra
> Como los anillos construidos en Saturno
> Habrá ciudades grandes como un país
> Gigantescas ciudades del porvenir
> En donde el hombre-hormiga será una cifra
> Un número que se mueve y sufre y baila
> (Un poco de amor a veces como un arpa hace olvidar la vida)
> Jardines de tomates y repollos
> Los parques públicos plantados de árboles frutales
> No hay carne que comer el planeta es estrecho
> Y las máquinas mataron el último animal
> Árboles frutales en todos los caminos
> Lo aprovechable sólo lo aprovechable
> Ah la hermosa vida que preparan las fábricas
> La horrible indiferencia de los astros sonrientes
> Refugio de la música
> Que huye de las manos de los últimos ciegos. (p. 380)

> [Someday after my death
> The world will seem small to everyone
> Continents will be planted in the seas
> There'll be islands in the skies
> There'll be a great metal bridge around the earth
> Like the rings constructed on Saturn
> There'll be cities as big as a country
> Gigantic cities of the future
> Where ant-man will be a cipher
> A number that moves and suffers and dances
> (With a little love sometimes like a harp that makes you forget about life)
> Gardens of tomatoes and cabbages
> Public parks planted with fruit trees
> There's no meat to eat space is tight
> And the machines killed the last animal

Fruit trees all along the roads
The profitable only the profitable
Oh the beautiful life the factories create
The horrible indifference of smiling stars
The shelter of music
That escapes from the hands of the last blind men] (p. 41)[6]

Verses by García Lorca, a contemporary of Huidobro, are also notable. In "La aurora" ["Dawn"] in *Poeta en Nueva York* [*Poet in New York*] (1929–1930), Lorca deals with an issue similar to the one mentioned in the verses of *Altazor*, the distance between the materiality of capital and the affective needs and intangible values so essential to human beings:

> La aurora de Nueva York tiene
> cuatro columnas de cieno
> y un huracán de negras palomas
> que chapotean las aguas podridas.
> La aurora de Nueva York gime
> por las inmensas escaleras
> buscando entre las aristas
> nardos de angustia dibujada.
> La aurora llega y nadie la recibe en su boca
> porque allí no hay mañana ni esperanza posible:
> A veces las monedas en enjambres furiosos
> taladran y devoran abandonados niños.
> Los primeros que salen comprenden con sus huesos
> que no habrá paraíso ni amores deshojados;
> saben que van al cieno de números y leyes
> a los juegos sin arte, a sudores sin fruto.
> La luz es sepultada por cadenas y ruidos
> en impúdico reto de ciencia sin raíces.
> Por los barrios hay gentes que vacilan insomnes
> como recién salidas de un naufragio de sangre. (pp. 119–120)

> [Dawn in New York has
> four columns of mire
> and a hurricane of black pigeons
> splashing in the putrid waters.
>
> Dawn in New York groans
> on enormous fire escapes

---

6   Huidobro, Vicente: *Obras completas. Tomo I*. Zig-Zag: Santiago de Chile 1964; Huidobro, Vicente: Weinberger, Eliot (trans.): *Altazor, or, A Voyage in a Parachute (1919): a Poem in VII Cantos*. Graywolf Press: Minneapolis 1988.

searching between the angles
for spikenards of drafted anguish.

Dawn arrives and no one receives it in his mouth
because morning and hope are impossible there:
sometimes the furious swarming coins
penetrate like drills and devour abandoned children.

Those who go out early know in their bones
there will be no paradise or loves that bloom and die:
they know they will be mired in numbers and laws,
in mindless games, in fruitless labors.

The light is buried under chains and noises
in an impudent challenge to rootless science.
and crowds stagger sleeplessly through the boroughs
as if they had just escaped a shipwreck of blood] (p. 11)[7]

Behind Huidobro, Lorca, and Neruda is not an indigenous worldview but the desire for a calm Earth, a desire associated with the imaginary of a sparsely populated paradise, the subject of the *locus amoenus*, or the discourse of *contempt of the court and praise for the village*; indeed, these are imagery and clichés that, given their classical sources and literary expressions in Judeo-Christian roots, would be, according to Paredes and McLean, incapable of serving the environmental cause. Nonetheless, there is agreement that these topics, however Western, praise spaces of calmness, slowness, and reflection that concur very well with the development of an ecological thought and presume a potential alternative to the system to which they belong, to productivism, speed, and consumption. From within the system, Huidobro, Lorca, and Neruda face the capitalist imaginary and, in doing so, approach the imaginary of the exploited, be they New York's black population or Latin America's original peoples, at least if we agree with Martínez Alier that

> las compañías transnacionales prefieren el lenguaje costo-beneficio [y los pobres suelen ser más versátiles y acuden] al lenguaje de lo sagrado, o al lenguaje del valor de los ecosistemas o paisajes. Frente al lenguaje costo-beneficio del mercado hay lenguajes completamente fuera del mercado, por ejemplo, los valores ecológicos de los ecosistemas (en términos de producción de biomasa, o en términos de riqueza de especies), el respeto

---

[7] García Lorca, Federico: *Antología poética*. Ediciones Orbis; Barcelona 1982; García Lorca, Federico: White, S. F. / Simon, G. (trans.): *Poet in New York*. Farrar Straus Giroux: New York 1988.

a su carácter sagrado, la necesidad ineludible de la subsistencia humana, los derechos de los animales, la dignidad de la vida humana, la demanda de seguridad alimentaria y ambiental, la defensa de la identidad cultural y los derechos territoriales indígenas, el valor estético de los paisajes, el valor de los derechos humanos, la lucha contra el racismo. (p. 332)[8]

[transnational companies prefer cost-benefit language [and the poor tend to be more versatile and attend] to the language of the sacred, or to the language of the value of ecosystems or landscapes. Faced with the market's cost-benefit language, there are languages completely outside the market, for example, the ecological values of ecosystems (in terms of biomass production, or in terms of species richness), respect for their sacred nature, the unquestionable need for human subsistence, the rights of animals, the dignity of human life, the demand for food and environmental security, the defense of cultural identity and indigenous territorial rights, the aesthetic value of landscapes, the value of human rights, the fight against racism].

In this regard, it would be pertinent to mention a beautiful passage from the fourth book of *Principles of Political Economy* (1848) in which John Stuart Mill, a neoclassical economist who, in full nineteenth century industrial development and in the era of blind faith in science and technology, contemplates the need to achieve a steady state. It is not surprising that George Sessions frames Stuart Mill in what he considers the three great historical opportunities that the West has not taken advantage of to abandon the path of "anthropocentric deviation" and return to "pre-Socratic ecocentrism": in the Middle Ages, the thinking of Jewish philosopher Maimonides and Saint Francis of Assisi; in the seventeenth century, Spinoza's pantheistic metaphysics; and, in the nineteenth century, the philosophy of Parson Thomas Malthus, John Stuart Mill and, together with them, Henry David Thoreau, George Perkins Marsh, John Muir and George Santayana, and a series of authors that, extending into the twentieth century, include Bertrand Russell, Aldous Huxley and Arne Naess, among others. According to Sessions, the failure of Saint Francis allowed the following:

> The anthropocentrism of the medieval Christian worldview, perpetuated in the philosophical systems of Bacon, Descartes and Leibniz, was to combine with, and be reinforced by, Renaissance anthropocentric humanism [...and] Renaissance humanism was to continue [...] with Karl Marx, John Dewey, and the humanistic existentialism of Jean-Paul Sartre. (p. 161)[9]

---

8   Martínez Alier, Joan: "Justicia ambiental, sustentabilidad y valoración". In: González de Molina, Manuel / Martínez Alier, Joan (eds.): *Naturaleza transformada*. Icaria: Barcelona 2001, pp. 289–336.
9   Sessions, George (ed.): *Deep Ecology for the 21st Century. Readings on the Philosophy and Practice of the New Environmentalism.*: Shambhala: Boston and London 1995.

Although modern science permitted the rescue of the non-anthropocentric pre-Socratic cosmology, first in astronomy with heliocentrism, the infinity of the universe, and cosmic evolution, and then in biology with Darwinian evolution, it failed to eradicate the anthropocentric matrix of Christianity and humanism that inspire its purpose to dominate and conquer Nature (Sessions, p. 162).

To the non-anthropocentric opportunities of authors and ideas alluded to by Sessions, one could add the decline of the writings of the Club of Rome, the bioeconomy of Nicholas Georgescu-Roegen in *The Entropy Law and the Economic Process* (1971), the criticism of the failure of globalization from authors such as Arturo Escobar, Gandhi's reflection on the practice of the simple life in *Indian Home Rule*, Günther Anders and *Die Antiquiertheit des Menschen* (1956), Hannah Arendt and *The Human Condition* (1958), the degrowth theory developed in the 1990s by Serge Latouche, Vincent Cheynet or François Schneider, and French publications such as *Silence* magazine, the weekly newspaper *La Décroissance*, or the book *Objective Decrease* (2003). Worth noting are The Institute of Economic Studies for the Sustainable Growth that was created in 2003 and of which Serge Latouche is president, and the first International Conference on Economic Degrowth for Ecological Sustainability and Social Equity that was held in Paris in 2008. Notably, this list includes Carlo Petrini's *Slow Food* (1986) movement which, of course, goes far beyond what its name suggests, and also the emergence and development of animal ethics from the criticism that Richard Ryder makes about the mistreatment of laboratory animals in *Victims of Science* (1975).[10]

In the aforementioned passage of the fourth book of *Principles of Political Economy* (1848), John Stuart Mill affirms:

> There is room in the world, no doubt, and even in old countries, for a great increase of population, supposing the arts of life to go on improving, and capital to increase. But even if innocuous, I confess I see very little reason for desiring it. The density of population necessary to enable mankind to obtain, in the greatest degree, all the advantages both of cooperation and of social intercourse, has, in all the most populous countries, been attained. If the earth must lose that great portion of its pleasantness which it owes

---

10 See Marrero Henríquez, José Manuel: "Animalismo y ecología: sobre perros parlantes y otras formas literarias de representación animal". *Castilla. Estudios de Literatura* 8, 2017, pp. 258–307. In this article, Marrero Henríquez studies, in light of contemporary environmentalist and animalist sensibilities, diverse forms of animal representation in historical samples of Spanish literature, from passages from songbooks to poems by authors active in the present moment. See also Marrero Henríquez, José Manuel: "Ética animal en *Coloquio de los perros*". *Ocnos* 17.3 (2018): 86–94.

to things that the unlimited increase of wealth and population would extirpate from it, for the mere purpose of enabling it to support a larger but not a better or a happier population, I sincerely hope, for the sake of posterity, that they will be content to be stationary, long before necessity compels them to it.

It is scarcely necessary to remark that a stationary condition of capital and population implies no stationary state of human improvement. Even the industrial arts might be as earnestly and as successfully cultivated, with this sole difference, that instead of serving no purpose but the increase of wealth, industrial improvements would produce their legitimate effect, that of abridging labor. Hitherto it is questionable if all the mechanical inventions yet made have lightened the day's toil of any human being. They have enabled a greater population to live the same life of drudgery and imprisonment, and an increased number of manufacturers and others to make fortunes. (pp. 594–595)[11]

Both the literary imagery of the quoted texts from Huidobro, Lorca, and Neruda's poetry and political economist Stuart Mill's stationary state defend nature against the gross exploitation of its resources and base themselves on a pristine and congenial nature, affected by the topics of solitude in tranquility, the pleasant place, the Garden of Eden, topicalized nature, and an attitude of belligerent refusal to equate well-being with the productivity and accumulation that are born within the heart of the consumer society. If we compare the ode "To Retired Life" by Spanish Renaissance poet Fray Luis de León to the poetry of Huidobro, Lorca, and Neruda and Stuart Mill's essay on political economy, far from supposing an idealized isolation of reality related to the perfect village of the *contempt of the court and praise of the village*, the pleasant garden promotes in the reader a responsive attitude to the accelerated pace of the city and the voracious model of production and consumption that sustains the system of contemporary life.

Fray Luis de León observes:

> Del monte en la ladera,
> por mi mano plantado tengo un huerto,
> que con la primavera,
> de bella flor cubierto,
> ya muestra en esperanza el fruto cierto;
>
> y como codiciosa
> por ver y acrecentar su hermosura,
> desde la cumbre airosa
> una fontana pura

---

11 Stuart Mill, John: "Stationary state of wealth and population dreaded by some writers, but not in itself undesirable". *Principles of Political Economy*. D. Appleton and Company: New York 1885, pp. 592–596.

hasta llegar corriendo se apresura;

y luego, sosegada,
el paso entre los árboles torciendo,
el suelo, de pasada,
de verdura vistiendo
y con diversas flores va esparciendo.

El aire el huerto orea
y ofrece mil olores al sentido;
los árboles menea
con un manso ruïdo
que del oro y del cetro pone olvido. [12]

[Upon the bare hillside
An orchard I have made with my own hand,
That in the sweet Springtide
All in fair flower doth stand
And promise sure of fruit shows through the land.

And, as though swift it strove
To see and to increase that loveliness,
From the clear ridge above
A stream pure, weariless
Hurrying to reach that ground doth onward press;

And straightway in repose
Its course it winds there tree and tree between,
And ever as it goes
The earth decks with new green
And with gay wealth of flowers spreads the scene.

The air in gentle breeze
A myriad scents for my delight distils,
It moves among the trees
With a soft sound that fills
The mind, and thought of gold or scepter kills]. (p. 171)

---

12 León, Fray Luis de: "Oda a la vida retirada", from http://www.dim.uchile.cl/~anmoreir/escritos/siglo_oro/fray.html#p1; León, Fray Luis de: Bell, Audrey F. G. (trans.): "To Retired Life". In: Turnbull, Eleanor Lairelle. *Ten Centuries of Spanish Poetry*. The Johns Hopkins Press: Baltimore 1969.

Complementary Differences

It is evident that nature's imaginaries are exactly that: imaginary and, consequently, cultural constructions. Precisely for this reason, they can be understood as autonomous forms alien to the natural, historical, and social conflicts that may surround them and also closely linked to specific environments and particular circumstances. With this dichotomy at hand, the literary imaginary of Western cultures would be condemned, locked in a tradition of quasi-solipsistic motifs and topics, while on the other, the imaginaries of the primitive cultures of America would be framed in the contexts of a primary oral tradition close to the conditions that nurture their wording. Nevertheless, despite the clarity of the outline, the oppositions, although being useful at times, reveal their inadequacies in such a way that, in the end, it is binary thought itself that does not give an account of an ecosystem as complex as that of hispanism and its literatures, an ecosystem susceptible to being reordered (even renamed: Indo-Hispanism? Hispano-Indianism?) to accommodate those literatures that originate from the traditions of the original peoples of America to those that come from Europe and are developed in hybrid cultural spaces.

As soon as we look closely at the oppositions that separate cultures of indigenous and European traditions, or Latin American versus Anglo-Saxon culture, we see confrontation as a strategy of distinction, insufficient and encouraging disputes. How does Christianity, which Rodó uses to vindicate Latin American spiritualism, not serve Max Weber's explanation of Anglo-Saxon materialism? In Sarmiento's civilization, there is barbarism and, in the barbaric, civilization; the Pachamama's anger is not only perceived in signs from nature but also in satellite observation of the withdrawal of the polar ice; the economy that Christian Felber bases on the European democratic constitutions seeks a common good that can also be found in the *sumak kawsay*, in the indigenous land uses that José Carlos Mariátegui defends, in Leonardo Boff's ecotheology, in Michel Serres's natural contract, in Enrique Dussel's biopolitics, and in the ecological doctrine of Pope Francis.[13]

---

13 The common good in Mariátegui is found in communal indigenism against feudal colonialism; in Serres, in the reorientation towards the nature of the social contract; in Boff, in the exacerbation of a planetary consciousness and of a cosmic democracy characterized by liberation by kindness, an awareness capable of remedying the problem of science as a standstill and of giving an alternative to the exploitative reading of Genesis; and, in Dussel, in the need to resolve the ethical issues that arise before the power that the human being has been acquiring over life, water, air, plants, embryos, stem cells, and euthanasia. In *sumak kawsay* and the encyclical letter *Laudato Si* of Pope Francis, the common good necessarily passes through the ecological respect of Creation.

As the proverb suggests, better a bad agreement than a good fight. Indeed, it is better to think of disagreements in terms of complementary differences, possible collaboration, and mutual enrichment than in confrontation. After all, the basis of all cultures, European or indigenous, is common nature, and the ultimate purpose of all cultures should be the well-being of its peoples. Martí envisioned it in *Our America* (1891), concluding that "there is no battle between civilization and barbarism, but between false erudition and nature" when he stated the following:

> el buen gobernante en América no es el que sabe cómo se gobierna el alemán o el francés, sino el que sabe con qué elementos está hecho su país, y cómo puede ir guiándolos en junto, para llegar, por métodos e instituciones nacidas del país mismo, a aquel estado apetecible donde cada hombre se conoce y ejerce, y disfrutan todos de la abundancia que la Naturaleza puso para todos en el pueblo que fecundan con su trabajo y defienden con sus vidas. El gobierno ha de nacer del país. El espíritu del gobierno ha de ser el del país. La forma de gobierno ha de avenirse a la constitución propia del país. El gobierno no es más que el equilibrio de los elementos naturales del país. Por eso el libro importado ha sido vencido en América por el hombre natural. Los hombres naturales han vencido a los letrados artificiales. El mestizo autóctono ha vencido al criollo exótico. No hay batalla entre la civilización y la barbarie, sino entre la falsa erudición y la naturaleza. (p. 17)[14]

> [the good governor in America is not one who knows how government is conducted in France or Germany, but who knows the elements of which his country is composed and how they can be marshaled so that by methods and institutions native to the country the desirable state may be attained wherein every man realizes himself, and all share in the abundance that Nature bestowed for the common benefit on the nation they enrich with their labor and defend with their lives. The government must be the child of the country. The spirit of the government must be the same as that of the country. The form of the government must conform to the natural constitution of the country. Good government is nothing more than the true balance between the natural elements of the nation. For that reason, the foreign book has been conquered in America by the natural man. The natural men have vanquished the artificial, lettered men. The native-born half-breed has vanquished the exotic Creole. The struggle is not between barbarity and civilization, but between false erudition and nature]. (p. 141)

Nevertheless, when an attempt is made to safeguard rights and preserve a people's culture in a legitimate manner, it is difficult to avoid self-definition by opposition, thereby moving the effort forward with renewed force. This concept

---

14 Martí, José: *Nuestra América*. Vitier, Cintio (ed.). Universidad de Guadalajara: Guadalajara 2002, from http://www.cucsh.udg.mx/cmarti/sites/default/files/nuestraa.pdf; Martí, José: Onís, Juan de (trans.): *The America of José Martí. Selected Writings*. Noonday Press: New York 1954.

is demonstrated in an interview by Joseph M. Pierce with Arturo Arias and Luis Cárcamo Huechante in which, because of the variety of linguistic and cultural options of the poets of the literatures and languages of the native peoples of America, the culture of the modern nation-state proves to be an enemy of the regional cultures of the original peoples of America:

> las literaturas indígenas representan una descolonización epistémica que articula nuevas geopolíticas del conocimiento, las cuales problematizan esa discursividad eurocéntrica que nos ha normado como académicos. [...En gran medida esas culturas] tienen mucho que aportar para la salvación del planeta en estos tiempos de recalentamiento ambiental y de crisis globales donde pululan los estados fallidos y el dominio de economías ilegales como las del narcotráfico o bien de la trata de seres humanos. (p. 5) [15]

> [indigenous literatures represent an epistemic decolonization that articulates new geopolitics of knowledge, problematizing the Eurocentric discursivity that has regulated us as academics. [... To a great extent, these cultures] have much to contribute to the salvation of the planet in these times of environmental warming and global crises in which the failed states and the domination of illegal economies, such as those of drug trafficking or trafficking in human beings, swarm].

As gathered from the words of Arturo Arias, it is true that the Latin American nation-state has subjugated part of the culture of the peoples that constitute it; however, as derived from the words of Arias himself, it is also true that a successful nation-state does not support drug trafficking or trafficking in human beings. In its constitution, a successful nation-state must include respect for the plurality of rights and the cultures that make up its citizenry. Although indigenous people and the modern nation-state are historical enemies, it is worth pointing out that they may end up finding a space for collaboration and mutual recognition. Among other things, a truly effective nation-state must welcome and give appropriate accommodation to the different cultures and peoples that comprise it, and it has the duty to harmonize their coexistence. A nation-state is a failed state and perverts its own *raison d'etre* when students, anti-corruption prosecutors, and indigenous environmental activists are assassinated with impunity, when incarcerations without legal or procedural guarantees occur, or when violence arises from historical grievances that remain pending resolution, as was the case during the January 2018 visit of Pope Francis to Temuco, capital of Araucanía.[16]

---

15 Pierce, J. M,: "Entrevista a Arturo Arias y Luis Cárcamo Huechante". *Pterodáctilo* 9, 2010, pp. 2–8, from. http://pterodactilo.com/numero9/files/2010/11/Pterodactilo_09_DossierEspecial.pdf.
16 On the incorporation of indigenous cultures and the notion of *sumak kawsay* in the national constitutions of Ecuador and Bolivia and on the relationships of their worldviews

Hispanoindigenism, Indohispanism

It is not a question of returning to Paredes and McLean's typology that opposes hispanism to indigenism, nor of removing the controversy with which, in 1946, hispanists and indigenists clashed over the exhumation of the bones of Hernán Cortés and then again over the bones of Cuauhtémoc exhumed on September 26, 1949; rather, the objective is to recontextualize those conflicts to emphasize that one is the soil that guards the bones of both and that this soil is identifiable with the global nature of the ecological crisis that affects everyone.[17] This common ground, both European and American, calls for the formulation of a poetics that responds to the challenges of the Anthropocene era from the Humanities and is capable of carrying out its critical activity by reading literary expressions with wisdom and acuity for, if they were only considered to be opposites by virtue of their Spanish-European or indigenous affiliation, they could never be understood by virtue of the ecological interests they share.

As Iván Carrasco M. affirms, it is true that Spanish-American literature "is a textual group governed by the conceptualization and norms of European origin" (p. 176), but it is also true that this conceptualization can and must be modified. In fact, Carrasco himself, in his definition of Chilean ethno-cultural poetry in the context of globalization, points to the necessary factors for such modification:

> [la poesía etnocultural es un] tipo artístico de textualidad bilingüe o plurilingüe, fundada en la experiencia de la interacción de grupos étnicos portadores de culturas, tradiciones artísticas, lingüísticas y textuales diferenciadas, que confluyen en el marco de una sociedad global donde comparten formas de vida, espacios, acontecimientos y experiencias. (p. 177)[18]

---

with pre-Socratic thought, see Heyd, Thomas: "The Natural Contract: Old Word vs. New World Perspectives". In: Marrero Henríquez, José Manuel (ed.): *Transatlantic Landscapes. Environmental Awareness, Literature and the Arts.* UAH: Alcalá de Henares: 2016, pp. 71–84. On the relationship of *sumak kawsay* to contemporary environmental movements of Mediterranean Europe such as "Slow Food" or the theories of degrowth, see Prádanos, Luis I.: "Decolonizing the North, Decolonizing the South: De-growth, Post-development, and Their Cultural Representations in Spain and Latin America". *Transatlantic Landscapes. Environmental Awareness, Literature and the Arts.* In: Marrero Henríquez, José Manuel (ed.): UAH: Alcalá de Henares 2016, pp. 49–70.

17 On the controversy of the authentic bones of Cortés and Cuauhtémoc, see López Portillo, Felícitas T.: "Hispanismo e indigenismo: sobre la polémica de los (verdaderos) huesos de Cortés y Cuauhtémoc". *Revista de la Universidad de México* 527, 1994, pp. 22–29.

18 Carrasco, Iván M. "La poesía etnocultural en el contexto de la globalización". *Revista de Crítica Latinoamericana* 58, 2003, pp. 175–192.

[(ethnocultural poetry is an) artistic type of bilingual or plurilingual textuality, based on the experience of the interaction of ethnic groups carrying differentiated cultures, artistic, linguistic and textual traditions that come together in the framework of a global society in which they share forms of life, spaces, events and experiences].

Iván Carrasco M. journeys through Chilean ethno-cultural poetry, from the initiators Luis Vulliamy (*Los rayos no caen sobre la yerba*, 1963) and Sebastián Queupul Quintremil (*Poemas mapuches en castellano*, 1966), and notes that, in general, the Chilean ethnocultural literature is based on the mixture of traditions, as is the case with the "Mapuche project",

> que tiene como objetivo crear una poesía intercultural que recupere la memoria ancestral de las comunidades nativas y está constituido por escritores de origen y cultura mapuche que han transformado su tradición epeu, ül, nütra, konew, en escritura regida por las normas de la literatura europea a partir de modelos chilenos e hispanoamericanos. Al hacerlo, han mezclado su expresión en mapudungun con la textualidad en español para redescubrir sus tradiciones, redefinir su identidad personal y enseñarnos que ninguna sociedad existe en el vacío y el aislamiento, sino en interacción con otras. (p. 184) [19]

[that aims to create an intercultural poetry that recovers the ancestral memory of the native communities and is constituted by writers of Mapuche origin and culture who have transformed their tradition epeu, ül, nütra, konew, into writing governed by the rules of European literature from Chilean and Hispanic American models. In doing so, they have mixed their expression in Mapudungun with textuality in Spanish to rediscover their traditions, redefine their personal identity and teach us that no society exists in emptiness and isolation, but in interaction with others].

Beyond the historical conflicts that remain to be resolved, the contact of ancient beings and ethnic presences with recent history highlights the great conflict of the present and immediate future that threatens all: the progress that endangers the nature that embodies the root of being, and not only of the Mapuche or Chilote being, but that of the Spanish-European being and of the human being in general. Sonia Caicheo's Chilote verses can stir the emotions of those Spaniards and Europeans who are as dissatisfied with the paths taken by progress as they are concerned about the prospects that scientists studying climate change put

---

19 Ivan Carrasco M.'s approach includes poets linked to the Mapuche culture such as Eric Troncoso, Clemente Riedemann, Juan Pablo Riveros, Pedro Alonzo Retamal, Elicura Chihuailaf, Lorenzo Aillapán, José Santos Lincomán, Jaime Luis Huenún, Bernardo Colipán, Rayen Kvyeh, Faumelisa Manquelipán, Adriana Pinda, Maribel Mora, Jacqueline Canihuán, Kelv Lihuen Tranamil and Chilote island poets such as Carlos Trujillo, Rosabetty Muñoz, Varsovia Viveros, Sergio Mansilla, Nelson Torres, Mario Contreras and Mario García.

on the table again and again: "El progreso llegó a la isla. El progreso / ¿Por qué estaremos, cada día, más tristes"? (Carrasco, p. 188) ["Progress came to the island. Progress / Why will we be, every day, sadder?"]. Similarly, the verses of Sergio Mansilla, in which the mythical character of a place confronts the present that threatens its qualities, could well move from the austral zone to any overexploited, Mediterranean, Atlantic or Caribbean beach, whose paradisiacal characteristics are threatened by industry in general or by the tourism industry in particular:

> En medio de la niebla
> oímos
> el murmurar de las playas ahora empobrecidas
> saqueadas, cerradas con alambre de púas
> por Transnacionales
> [...]
> Remen, remen, boteros,
> para que no se termine la eternidad. (Carrasco, p. 189).
>
> [In the middle of the fog
> we hear
> the murmuring of beaches now impoverished
> looted, closed with barbed wire
> by Transnational
> [...]
> Row, row, boatmen,
> so that eternity does not end].

From the twentieth century onward, it is advisable to defend the ecological values that cultures treasure, some with oral and indigenous roots, some with literary and European roots, all to some extent hybrid. Seen from this light, the texts previously quoted by Huidobro, Lorca, and Neruda, including the verses of Fray Luis de León's "To Retired Life", find a kinship of familiarity with those of the Mapuche poetry of Sonia Caiceo and Sergio Mansilla because, in the poetry of one to another, the words breathe and poetically recognize themselves within nature, which behind the distinct characteristics of their cultures, breathes and demands care.

Beauty in Nature and Culture

In the anthropology of Lévi-Strauss, Derrida emphasized the binary oppositions of structuralism, whose value is not truth, but methodology. In the Anthropocene era, this approach must give way to relational thinking that allows for the acceptance of reality's multipolarity that, as Bart Kosko points out, is "fuzzy" and consists of innumerable nuances of gray. If culture is part of nature, the hypothesis that the written forms of Hispanic literary texts preserve the desire to grasp the immediate knowledge of the environment, which Walter Ong studies in primary orality or Bruno Latour describes in the paradoxical movement of scientific investigation by which one moves away from reality to approach it, points to a commendable goal. In Hispanic literatures, Latin American or European, those akin to indigenous cosmogonies, Christian lineage, or declaredly atheist, or of oral, written or digitally virtual origin, Anthropocene ecocriticism has the important generic and transboundary task of highlighting the complementarity of complex artistic processes and the vital breath they share and through which, in literary writing, even nature breathes.

As a theory of ecological criticism, the poetics of breathing understands culture as a derivative of nature, and, as Martí himself pointed out, the opposition between nature, indigenism, orality, countryside, barbarism, on the one hand, and culture, Europeanism, writing, city, civilization, on the other, refers to the opposition of civilization / barbarism, which is an incorrect opposition. The beauty of literary imaginaries, their potential breath, will be the object of study in an ecocriticism inspired by the poetics of breathing, and beauty will be understood in any tradition, oral or written, indigenous or European, as a sophisticated cultural manifestation of the natural instinct for survival derived from the impulse to know the medium and make it intelligible to the breath.

The understanding of beauty is the result of the instinct of life and, at the same time, a basic sign of culture. Beauty is found everywhere in natural life, which flows calmly in cycles and repetitions, in biological forms characterized by patterns of order in time and space. The beauty of the fractality of plants and the vascular system of animals is omnipresent in the biology of life because biology needs regularities to successfully achieve life's objective. The encounter of these regularities, sometimes difficult to grasp, depends on aesthetic experience and scientific knowledge. As Jorge Wagensberg states,

> [...The] mind has evolved submerged in nature. We could better say that the mind has co-evolved with nature within it. Therefore it may not be going too far to assure that the ratio of order that provokes overall mental pleasure is precisely the ratio of natural order. [...] In the essence of mental pleasure there is perhaps the same quantity of rhythm and harmony exhibited in nature. In other words, whether something is too predictable or too unpredictable can perhaps be measured with respect to this reference, the natural order of things. This means that mental pleasure is a common property for all humans individuals, and who knows for how many other animal

species. It is a universal capacity because it is associated with physics, physiology and the psychology of the perception of uncertainty in the world that surrounds us. (p. 190)[20]

Beauty is such a fundamental feature of life that the predisposition to understand it has reached far beyond being a natural survival instinct. Permeating every aspect of existence, beauty is a foundational pillar of culture. The rhythm and harmony are not only everywhere in the natural world, they are also, as Wagensberg affirms, the foundation for the understanding of the intelligibility of things:

> Time and space are the *a priori* concepts with which we construct knowledge. Rhythm and harmony, then, form the basic structure of intelligibility. Rhythm and harmony are also, in their presence or absence, the essential concepts of beauty. […] Beauty not only precedes intelligibility. Beauty creates a predisposition for intelligibility. (p. 185)

The historical process that has made human expression an artistic inquiry into a complex cultural context determined by grammar, literary genres, rhetorical techniques, poetic speculations, schools, universities and other academic institutions does not cancel the primitive impulse of human expression to give a finished account of the surrounding environment within a context of primary orality. On the contrary, literary conventions and the cultural distancing of nature contribute to increase the desire for knowledge and the sense of belonging to a place, whether it be the universe, planet Earth, nation, region, city, neighborhood, or the rustic family farm. When written words breathe, the aesthetic commitment contributes to the environmental awareness and approach to the surrounding environment.

Written sophistication can thus be understood as the cultural procedure that seeks a relationship with the world similar to that of primary oral cultures. Upon the arrival of literary literacy, and even at the point of arrival of digital and multimedia literacy that nourishes cyber-poetry, criticism should not forget the itinerary of orality towards the literate society, at least if the criticism aspires to be ecologically relevant.[21]

---

20 Wagensberg, Jorge: "Beauty and Intelligibility". In: Marx, Mina / Sevilla, Manuel (eds.): *Fernando Casás. Arqueología del no-lugar*. Círculo de Bellas Artes: Madrid 2004, pp. 25–38.
21 With poetic clarity, David Abram has explained the link that unites the complexity of artistic and scientific languages, which have their own internal coherence, with primary orality's immediate contact with the environment when he states: "It is a commonplace to observe that today the perceived world is everywhere filtered and transformed by technology, altered by the countless tools that interpose themselves between our senses and the early sensuous. It is less common to suggest that there's a wildness that

Poetics of Breathing

The intimate relationship of culture and nature, of writing and orality, is revealed in beauty, and it is beauty's relevance in nature and culture that is why Anthropocene ecocriticism is "aerial", necessarily transcultural, transborder, and transnational. Words written in a context of literate culture share with those coming from a context of orality the common impulse to unravel the sense of life's rhythm and harmony. A literary work not only indicates its language and cultural conventions with its internal disciplinary coherence, but also responds to the instinct to give an explanation of the world as immediate as that which corresponds to an onomatopoeia or a primary oral sound with the reality that is there outside.[22]

In the last chapter of *The Air and the Dreams*, entitled "The Silent Declamation", Gaston Bachelard describes poetry's primary status in terms that recall the onomatopoeic explanation of the world and the significant sigh before nature:

> In its simple, natural, and primitive form, far from any aesthetic ambition and from any metaphysics, poetry is the happiness of a sigh, the evident joy of breathing [...] In strength and in tenderness, in choleric and temperate poetry, there is a purposively oriented economy of breathing in action, there is a suitable administration of the air that speaks. At least poems with *good breathing* are that way, poems that are beautiful dynamic schemes of breathing. (p. 294)[23]

The interest in mimology for an ecocriticism that seeks to explain the processes by which words breathe does not rest in its scientific basis, but in the fascination it demonstrates for the possibility of establishing solid relationships between words and meanings and for the desire to overcome *la defaut des langues*.[24] The utopia of perfection embodied in mimology points to the natural instinct to perceive beauty and pursue the intelligibility of the things that are at the rhythmic and harmonic basis of the arts and sciences, from poetry to physics, from computational engineering to architecture. Not in vain, Thaïs E. Morgan appeals to

---

still reigns underneath all these mediations--that our animal senses, coevolved with the animate landscape, are still tuned to the many-voiced earth" (p. 264). Abram, David: *Becoming Animal: An Earthly Cosmology*. Pantheon Books: New York 2010.

22 For an outline of the poetics of breathing, see Skowronski, Ellen: "Words that Breathe. An Interview with José Manuel Marrero Henríquez". *Ecozon@. European Journal of Literature, Culture and Environment* 6(1), 2015, pp. 107–117.

23 Bachelard, Gaston: *El aire y los sueños*. Fondo de Cultura Económica: México 1986.

24 Borges approaches the subject of the insufficiencies of language with exquisite irony in "The Analytical language of John Wilkins".

Stéphane Mallarmé's famous phrase in his preface to the English translation of *Mimologiques: Voyage en Cratylie* by Gérard Genette:

> la defaut des langues, "the failing of natural languages" or "the defect of natural languages" [...to highlight that] the project of secondary Cratylism espoused by most of the philosophers, linguists, and literary writers mentioned in *Mimologics* is to "correct" the "defect" of proper and ordinary names (nouns) by inventing alternative, supposedly better—because more mimetic, more motivated, less arbitrary—sound symbolism, alphabets, writing systems, and so on. (xvi)[25]

Is this not what happens in the writing of poetry? Perhaps in poetry the arbitrariness of language and the literary resources of rhetoric and literary genres do not favor the need for the expression of primary orality? Seemingly affirming this idea, Dámaso Alonso decades ago pointed out the following:

> para Saussure el signo, o con otras palabras, la relación entre significante y significado, es siempre arbitraria. Bien, para nosotros, en poesía, hay siempre una relación motivada entre significante y significado. Éste es nuestro axioma de partida. Entendemos *poesía* en el sentido general del alemán *Dichtung*; y podemos añadir que la relación motivada entre significante y significado se hace más patente en la poesía en verso, especialmente en la poesía lírica o en la narrativa fuertemente caracterizada por su lirismo. (p. 32)[26]

> [for Saussure the sign, or with other words, the relation between signifier and signified, is always arbitrary. Well, for us, in poetry, there is always a motivated relationship between signifier and meaning. This is our starting axiom. We understand *poetry* in the general sense of the German *Dichtung*; and we can add that the motivated relationship between signifier and meaning becomes more evident in poetry in verse, especially in lyric poetry or in narrative strongly characterized by its lyricism].

In Ernesto Cardenal's verses, one finds a poetic example of cultures that, although at first glance seem alien to one another, result in being complementary, as they marry science with spiritual intuition and return the scientific truth to the sacred explanation that Martínez Alier sets against the productivist reason of capitalism:

> Acércate a esta roca junto al mar y mira:
> es casi enteramente espacio vacío
>   (mírala electrónicamente)
> es evanescente espuma toda ella
> como la espuma de mar que de las rocas nace y en las rocas se deshace.

---

25 Genette, Gérard: *Mimologics*. University of Nebraska Press: Lincoln 1995.
26 Alonso, Dámaso. *Poesía española. Ensayo de métodos y límites estilísticos. Garcilaso, Fray Luis de León, San Juan de la Cruz, Góngora, Lope de Vega, Quevedo.* Madrid: Gredos, 1971.Translations of Dámaso Alonso are my own.

Efímeras partículas que no están ni aquí ni allí,
yendo y viniendo al azar de las olas de un mar vacío.
Partículas que surgen de la nada y vuelven al olvido.
    Viajan del vacío al vacío.
[...]
    ¡El impalpable mundo palpable!
¿Es posible que la naturaleza sea tan absurda?
Se preguntaba Heisenberg en el parque de madrugada, en la incierta
luz danesa, tras las largas discusiones con Bohr.
[...]
    Las partículas insustanciales de la materia.
¿O será como decir: la materia espiritual del cosmos?
[...]
    Transparente como el espíritu
    la materia no es más
que espacio vacío. (pp. 295–296)

[Approach this rock by the sea, and look:
it's almost entirely empty space
    (look at it electronically)
it's evanescent foam, all of it,
like sea foam that's born from the rocks and on the rocks is undone.
ephemeral particles that are neither here nor there,
coming and going at the whim of the waves of an empty sea.
Particles which spring from nothing and return to oblivion.
    From emptiness to emptiness they travel.
[...]
    The impalpable palpable world!
Can nature possibly be so absurd?
Heisenberg asked himself in the park at dawn, in the wavering
Danish light, following his lengthy discussions with Bohr.
[...]
    The insubstantial particles of matter.
Or could it be like saying: the spiritual matter of the cosmos?
[...]
    Transparent like the spirit
        Matter is nothing more
than empty space
        and fields of energy. (pp. 274–277)[27]]

---

27 Cardenal, Ernesto. *Canto cósmico*. Editorial Trotta: Madrid 1992; Cardenal, Ernesto: Lyons, Jonathan (trans.): *Cosmic Canticle*. Curbstone Press: Evanston 2002.

The empty space of Ernesto Cardenal's verses, which is "transparent as the spirit" and transparent as matter, is the air breathed in the written words of *Cosmic Canticle*, allowing scientific inquiry in nature to identify with its aesthetic elaboration and spiritual understanding. Cardenal links the seemingly unlinkable by resorting to what is shared by scientific methodology, literate sophistication, and the spirituality of the oral cultures of the peoples of America. The words that are rhythmically breathed in *Cosmic Canticle* reflect the rhythm and harmony of nature and culture, and they demonstrate that the roots of scientific knowledge, aesthetic understanding, and spiritual perceptions of the universe are based on regularities in the time and space that shape nature. In the poetics of breathing, there is no place for scientific pride or aristocratic conception of art, for within it dominates the idea that criticism achieves life only when it inspires the air that is the root of the written words that breathe. Not in vain did Lorca's verses warn that light does not reach human beings when "Light is buried under chains and noises / in the impudent challenge of science without roots".

Bibliography

Abram, David: *Becoming Animal: An Earthly Cosmology*. Pantheon Books: New York 2010.

Alonso, Dámaso: *Poesía española. Ensayo de métodos y límites estilísticos. Garcilaso, Fray Luis de León, San Juan de la Cruz, Góngora, Lope de Vega, Quevedo*. Gredos: Madrid 1971.

Anders, Günther: *La obsolescencia del hombre*. Pre-Textos: Valencia 2011.

Arendt, Hannah: *The Human Condition*. The University of Chicago Press: Chicago 1958.

Bachelard, Gaston: *El aire y los sueños*. Fondo de Cultura Económica: México 1986.

Barbas-Rhoden, Laura: "Hacia una ecocrítica transnacional: aportes de la filosofía y crítica cultural latinoamericana a la práctica ecocrítica". *Revista de Crítica Literaria Latinoamericana* 79, 2014, pp. 79–98.

Binns, Niall: "¿Puro Chile es tu cielo azulado? Poesía ecologista en la *delgada patria* (Vicente Huidobro, Gabriela Mistral, Pablo Neruda y Nicanor Parra)". *Ixquic* 2, 2000, pp. 38–54.

Binns, Niall: "Poéticas para un mundo insostenible". In: Marrero Henríquez, José Manuel (eds.): *Literatura y sostenibilidad en la era del antropoceno*. Fundación MAPFRE-Guanarteme: Las Palmas de Gran Canaria 2011, pp. 59–76.

Boff, Leonardo: *Ecología: grito de la tierra, grito de los pobres*. Lumen: Buenos Aires 1996.

Borges, Jorge Luis: "El idioma analítico de John Wilkins". *Obras completas*. Emecé: Buenos Aires 1995, pp. 706–709.

Cardenal, Ernesto: *Canto cósmico*. Editorial Trotta: Madrid 1992.

Cardenal, Ernesto: Lyons, Jonathan (trans.): Curbstone Press: New York 2002.

Carrasco, Iván M.: "La poesía etnocultural en el contexto de la globalización". *Revista de Crítica Latinoamericana* 58, 2003, pp. 175–192.

Derrida, Jacques: "La estructura, el signo y el juego en el discurso de las ciencias humanas". *La escritura y la diferencia*. Anthropos: Barcelona 1989, pp. 383–401.

Dussel, Enrique: *Filosofía de la liberación*. Fondo de Cultura Económica: México 2011.

Escobar, Arturo: "Economics and the Space of Modernity: Tales of Market, Production and Labour". *Cultural Studies* 19(2), 2005a, pp. 139–175.

Escobar, Arturo: "El 'postdesarrollo' como concepto y práctica social". In: Mato, Daniel (coord.): *Políticas de economía, ambiente y sociedad en tiempos de globalización*. Universidad Central de Venezuela: Caracas 2005b, pp. 17–31.

Escobar, Arturo: "Visualización de una era postdesarrollo". *La invención del tercer mundo. Construcción y deconstrucción del desarrollo*. Editorial Norma: Bogotá 2006, pp. 397–424.

Felber, Christian: *La economía del bien común*. Deusto Ediciones: Bilbao 2012.

Flys Junquera, Carmen / Marrero Henríquez, José Manuel / Barella Bigal, Julia (eds.): *Ecocríticas*. Iberoamericana Vervuert: Madrid 2010.

Gandhi M. K.: *Indian Home Rule*. The International Printing Press: Natal 1910.

García, Mabel: "El proceso de retradicionalización cultural en la poesía mapuche actual: *Üi* de Adriana Paredes Pinda". *Revista Chilena de Literatura* 81, 2012, pp. 51–68.

García Lorca, Federico: *Antología poética*. Ediciones Orbis: Barcelona 1982.

García Lorca, Federico: White, S. F. / Simon, G. (trans.): *Poet in New York*. Farrar Straus Giroux: New York 1988.

Genette, Gérard: *Mimologics*. University of Nebraska Press: Lincoln 1995.

Georgescu-Roegen, Nicholas: *The Entropy Law and the Economic Process*. Harvard University Press: Cambridge 1971.

Heffes, Gisela: "Introducción. Para una ecocrítica latinoamericana: entre la postulación de un ecocentrismo crítico y la crítica a un antropocentrismo hegemónico". *Revista de Crítica Literaria Latinoamericana* 79, 2014, pp. 11–34.

Heyd, Thomas: "The Natural Contract: Old Word vs. New World Perspectives". In: Marrero Henríquez, José Manuel (ed.): *Transatlantic Landscapes*.

Environmental Awareness, Literature and the Arts. UAH: Alcalá de Henares 2016, pp. 71–84.

Huidobro, Vicente: *Obras completas. Tomo I*. Zig-Zag: Santiago de Chile 1964.

Huidobro, Vicente: Weinberger, Eliot (trans.): *Altazor, or, A Voyage in a Parachute (1919): a Poem in VII Cantos*. Graywolf Press: Minneapolis 1988.

Kosko, Bart: *Pensamiento borroso. La nueva ciencia de la lógica borrosa*. Crítica: Barcelona 1995

Latouche, Serge: *Faut-il refuser le développement?: Essai sur l'anti-économique du Tiers-monde*. Presses universitaires de France: Paris 1986

Latour, Bruno: *Pandora's Hope. Essays on the Reality of Science Studies*. Harvard University Press: Cambridge and London 1999.

León, Fray Luis de: "Oda a la vida retirada" from http://www.dim.uchile.cl/~anmoreir/escritos/siglo_oro/fray.html#p1.

León, Fray Luis de: Bell, Audrey F. G. (trans.): "To Retired Life". In: Turnbull, Eleanor Lairelle. *Ten Centuries of Spanish Poetry*. The Johns Hopkins Press: Baltimore 1969.

Lévi-Strauss, Claude: *El pensamiento salvaje*. Fondo de Cultura Económica: México 1997.

Lienlaf, Leonel: *Se ha despertado el ave de mi corazón*. Universitaria: Santiago 1989.

Lienlaf, Leonel: "3 poemas". *Pterodáctilo* 9, 2010, pp. 20–22, from http://pterodactilo.com/numero9/files/2010/11/Pterodactilo_09_DossierEspecial.pdf.

López Portillo, Felícitas T.: "Hispanismo e indigenismo: sobre la polémica de los (verdaderos) huesos de Cortés y Cuauhtémoc". *Revista de la Universidad de México* 527, 1994, pp. 22–29.

Mariátegui, José Carlos: *Siete ensayos de interpretación de la realidad peruana*. Linkgua: Barcelona 2012.

Marrero Henríquez, José Manuel (coord.): *Lecturas del paisaje*. Servicio de Publicaciones de la ULPGC: Las Palmas de Gran Canaria 2009.

Marrero Henríquez, José Manuel: "Ecocrítica e hispanismo". In: Flys Junquera, Carmen et al. (eds.): *Ecocríticas. Literatura y medio ambiente*. Iberoamericana Vervuert: Madrid 2010, pp. 193–218.

Marrero Henríquez, José Manuel (ed.): *Literatura y sostenibilidad en la era del antropoceno*. Las Palmas de Gran Canaria: Fundación MAPFRE-Guanarteme, 2011a.

Marrero Henríquez, José Manuel: "Literary Waters in a Dry Spain". *ISLE. Interdisciplinary Studies in Literature and Environment* 18.2 (2011b): 413–430.

Marrero Henríquez, José Manuel: "Pertinencia de la ecocrítica". *Revista de Crítica Literaria Latinoamericana* 79 (2014): 57–78.

Marrero Henríquez, José Manuel (ed.): *Transatlantic Landscapes. Environmental Awareness, Literature and the Arts*. Franklin Institute – UAH: Alcalá de Henares 2016.

Marrero Henríquez, José Manuel: "Ética animal en *Coloquio de los perros*". *Ocnos* 17.3 (2018): 86–94.

Martí, José: Onís, Juan de (trans.): *The America of José Martí. Selected Writings*. Noonday Press: New York 1954.

Martí, José: *Nuestra América*. Vitier, Cintio(ed.). Universidad de Guadalajara: Guadalajara 2002, from http://www.cucsh.udg.mx/cmarti/sites/default/files/nuestraa.pdf.

Martínez Alier, Joan: "Justicia ambiental, sustentabilidad y valoración". In: González de Molina, Manuel / Martínez Alier, Joan (eds.): *Naturaleza transformada*. Icaria: Barcelona 2001, pp. 289–336.

Meeker, Joseph W.: *The Comedy of Survival. Literary Ecology and a Play Ethic*. The University of Arizona Press: Tucson 1997.

Neruda, Pablo: O'Daly, William (trans.): *World's End*. Copper Canyon Press: Washington 2009.

Neruda, Pablo: Fin de mundo. Losada: Buenos Aires 1969.

Neruda, Pablo: *Obras completas. III. Pablo Neruda. De Arte de pájaros a El mar y las campanas. 1966–1973*. Loyola, Hernán (ed.). Galaxia Gutenberg y Círculo de Lectores: Barcelona 2000.

Ong, Walter: *Oralidad y escritura: tecnologías de la palabra*. Fondo de Cultura Económica: México 1987.

Papa Francisco: "Carta encíclica *Laudato Si* sobre el cuidado de la casa común", 2015, from http://www.connect4climate.org/images/uploads/papa-francesco_enciclica-laudato-si_ESPANOL.pdf.

Paredes, Jorge / McLean, Benjamin: "Hacia una tipología de la literatura ecologista en el mundo hispano". *Ixquic* 2, 2000, pp. 1–37.

Petrini, Carlo: *Bueno, limpio y justo: principios de una nueva gastronomía*. Ediciones Polifemo: Madrid 2007.

Pickering, Andrew / Guzik K. A.: "New Ontologies". In: Pickering, Andrew / Guzik K, A. (eds.): *The Mangle in Practice: Science, Society and Becoming*. Duke University Press: Durham 2008, pp. 1–14.

Pierce, J. M. "Entrevista a Arturo Arias y Luis Cárcamo Huechante". *Pterodáctilo* 9, 2010, pp. 2–8, from http://pterodactilo.com/numero9/files/2010/11/Pterodactilo_09_DossierEspecial.pdf.

Prádanos, Luis I.: "Decolonizing the North, Decolonizing the South: De-growth, Post-development, and Their Cultural Representations in Spain and Latin America". In: Marrero Henríquez, José Manuel (ed.): *Transatlantic Landscapes. Environmental Awareness, Literature and the Arts.* UAH: Alcalá de Henares 2016, pp. 49–70.

Sarmiento, Domingo Faustino: *Facundo. Civilización y barbarie.* Emecé: Buenos Aires 1999.

Serres, Michel: *El contrato natural.* Pre-Textos: Valencia 1991.

Sessions, George (Ed.): *Deep Ecology for the 21st Century. Readings on the Philosophy and Practice of the New Environmentalism.* Shambhala: Boston and London 1995.

Skowronski, Ellen: "Words that Breathe. An Interview with José Manuel Marrero Henríquez". *Ecozon@. European Journal of Literature, Culture and Environment* 6(1), 2015, pp. 107–117.

Stuart Mill, John: "Stationary state of wealth and population dreaded by some writers, but not in itself undesirable". *Principles of Political Economy.* D. Appleton and Company: New York 1885, pp. 592–596.

Tamames, Ramón: *La polémica sobre los límites al crecimiento.* Alianza: Madrid 1974.

Vásquez M., Paola (dir.): *Mapuche de ayer y de hoy.* 2007, from https://www.youtube.com/watch?v=rQlNq5gkPNE.

Wagensberg, Jorge: "Beauty and Intelligibility". In: Marx, Mina / Sevilla, Manuel (eds.): *Fernando Casás. Arqueología del no-lugar.* Círculo de Bellas Artes: Madrid 2004, pp. 25–38.

Weber, Max: *La ética protestante y el espíritu del capitalismo.* Alianza: Madrid 2012.

Jorge Marcone
Towards an Amazonian Environmental Humanities

Abstract: For more than a decade now, one of the preferred areas of study of the then emergent field of ecocriticism in Latin American literature has been the *novela de la tierra* and the *novela de la selva*, or the Regionalist novel, for short. A variety of serious studies has focused on two central themes. The first one is the intervention of international capitalism and the national state in the Amazon. In this case, ecocritical approaches have called our attention to the fact that, in the Regionalist novel, the Amazon is a site of neo-colonialism associated with the extractivist industries that started with the Rubber Boom in 1880. Secondly, the scholarship has extensively studied how in these fictions the Amazon and its denizens become the ultimate Other for the European or *criollo* traveler or entrepreneur stressed to the limit by the tropical environment. To a great extent, this ecocritical scholarship is an indirect comment, and even resistance, to the current scenario of a renewed extractivism in the Amazon. Nevertheless, the current resistance by indigenous peoples and small farmers on the ground to this extractivism sponsored or supported by the national states suggests new leads for further understanding Amazonian literatures. "Amazonian", for the purposes of this chapter, is understood very broadly: indigenous or nonindigenous narratives, written by scientific travelers or state officials; or focusing on subsistence farmers or poor immigrants in search of fortune. In sum, we suspend for a moment the distinction between literature written by "insiders" and "outsiders". The new ecocritical insights indirectly suggested by the popular environmentalisms themselves derive from the success of these movements in introducing into the public political debate different ontologies on the human and the nonhuman. As a matter of fact, the indigenous cosmologies invoked by the environmental movements have become the focus of philosophical debates on the status of the nonhuman as subjects and objects of politics. Amazonian cosmologies or ontologies are currently informing the efforts of the so-called new materialisms for imagining philosophical alternatives to the nature/society dichotomy, and to the naturalist or culturalist reductionism of human/nonhuman relations that this dichotomy supports. These efforts also include a revision of the ontological status of fictional, mythological, and imaginary beings in general, and therefore, of the notion of text in modern and postmodern literary theories. Instead of emphasizing that these beings are "mental" entities supported by language, the "new ontologies" experience these beings akin to the perception of invisible human and nonhuman beings in Amazonian shamanism. At the convergence of the impact of popular Amazonian environmentalism in national politics, and of the philosophical debates on the political ontologies implied in those social movements, emerges the need to ask now if the notions of human and nonhuman, and of text as a cultural construct, that so far have supported ecocritical readings of Amazonian literatures and film, are in line with Amazonian ontologies. Is it possible to set forth an Amazonian ecocritical theory that would not rely on a notion of text built upon the split between nature and society, human and nonhuman? Our thesis is that, following recent anthropological scholarship fostering the encounter of Amazonian ontologies with philosophy on the subjects and objects of politics, enunciations in any media can be thought of as entities originated in the interaction of human and nonhuman actors. Furthermore, the enunciations themselves can be thought of as more-than-human. An Amazonian ecocritical theory would help us explain and overcome the systematic absence of Amazonian texts in every national literary canon in the so-called Andean countries. And this exclusion happens in spite of the wide academic acceptance of transcultural or intercultural theories of literature and other artistic production. Thus, an Amazonian ecocritical theory would also mean involving the Humanities in the larger political experiment of building a common world in countries characterized by their cultural plurality and by the plurality of ontological worlds coexisting, or in conflict.

Keywords: Ecocritical Theory, Environmental Humanities, Amazonian Literature, Regionalist Novel

Old Stories for Current Times

The public debate in Latin America on indigenous or peasant resistance to extractivism is particularly unaware of the deep reflection regarding this kind of political economy that took place in *indigenista* literature, the *novela de la tierra*, and the *novela de la selva* of the first half of the twentieth century in Latin America. Let us examine, for instance, the *novela de la selva*. Even though the Amazon was strongly protected from human impact until the 1960s, due to remoteness from major settlements, narratives in this genre addressed, in ways still relevant nowadays, the impact of extractivism on local populations and environments, and on national politics as well. Nowadays, politicians, policymakers, government officials, and a divided public seem to be lost or indolent as to how to engage with indigenous values and practices. But we can also wonder about how much this inability to prevent or solve environmental conflicts is linked to a lack of knowledge or interest in the reflections on the impact of different waves of modernization that have emerged precisely at the intellectual core of the nine South American countries with sovereignty over the Amazon.

The *novela de la selva* is by now a genre very well studied in ecocriticism. These studies have focused on three central themes. The first one is the intervention of neo-colonialism associated with the extractivist industries that started with the Rubber Boom in the 1880s. Secondly, the scholarship has extensively studied how in these fictions the Amazon and its denizens, humans and nonhumans, become the ultimate Other for the European or *criollo* traveler or migrant stressed to the limit by the tropical environment. And, finally, we understand now these narratives are the seed of an awareness of not exactly an environmental degradation of planetary consequences but rather of crimes and cruelty done to humans and nonhumans that calls for an ethical and legal response on the part of all citizens of the nation-state.

Ecocritical studies (or, for this matter, environmental humanities, posthumanist or post-natural studies) are increasingly paying attention to current scenarios in which Andean and Amazonian popular environmentalisms against extractivism, indigenous or not, are having an impact on national politics in two aspects: first, in the development of notions of well-being alternative to understandings of such aspiration in conventional economic development or neo-liberalism; and secondly, with the idea that nonhuman "natural" entities could be, or must be, considered subjects of rights or political agents in such ways that they challenge the subject/object dichotomy in conventional political ontologies. In fact, it is well known that through Amazonianist anthropologists, these indigenous cosmologies are intervening in debates of political ontology that, in turn, want to shake up other fields.

Given this state of affairs, it is hard to draw attention to the *novela de la selva*. The *novela de la selva* was written by nonnative Latin American writers, "outsiders" to the Amazon; and these fictions, written according to the aesthetics of their time, are something very different from allegedly "true" Amazonian stories retold orally by native or mestizo shamans, or in writing by Amazon-born intellectuals. Despite these issues, I want to begin this chapter on Amazonian environmental humanities by returning to the *novela de la selva* once more in order to bring to the attention of an interdisciplinary or multidisciplinary audience the fact that these narratives are as relevant and insightful as ever to the diverse Amazonian environmental agendas and, in addition, to suggest a pending task in ecocritical studies regarding the ontological status of any Amazon-related text or cultural artifact at a time when we are just catching up with the direction in which we are heading as we take an "ontological turn" in the environmental humanities.

For those readers not familiar at all with the *novela de la selva*, let me attempt a brief definition. Among many other titles forgotten since the second half of the twentieth century, this genre includes significant canonical fictions of Latin America's literary tradition: the so-called cuentos misioneros (Uruguay, 1918–1924) by Horacio Quiroga, *La vorágine* (Colombia, 1924) by José Eustasio Rivera, *Canaima* (Venezuela, 1935) by Rómulo Gallegos, *Los pasos perdidos* (Venezuela, 1954) by Alejo Carpentier, and *La casa verde* (Peru, 1965), among other fictions, by Mario Vargas Llosa. In the *novela de la selva*, a modern, urban, colonial and male subject becomes disappointed with the city for not fulfilling his expectations of emancipation, self-realization, financial independence, and/ or artistic authenticity. As the subject flees the city, he progressively conceives of and attempts a "return to nature" in the wilderness of the tropical forest. It is in close interaction with a challenging tropical landscape where the subject deliberately seeks freedom, identity, and renewal. For the European and *criollo* travelers or migrants, the Amazon offers a diversity of experiences: a site where the prehistorical origins of the world live on, a laboratory of evolutionary history, or a refuge of ahistorical and universal truths about humans and nature. Nevertheless, this search for an alternative modern lifestyle, alternative to the city but also to extractivist activities (alternative development, we would call it now), usually ends in failure.

The failure of this pursuit of well-being and happiness is due to a combination of the following factors. (1) The tropical forest resists control by the technology transferred to the place, and therefore raises the issue of the lack of sound and reliable scientific and historical knowledge of the region. (2) Under extreme physical and psychological conditions, more than a few characters end up recognizing, against modernist principles, sometimes in fear or at least with suspicion of their own perception, personhood in the environment, the weather, plants and animals, bodies of water, and other invisible nonhumans. (3) The experience of the jungle proves to be interdependent of the social relationships that, in fact, made the trip and the settlement possible, but this revelation comes in the form of a nuisance. In this respect, Amazonian indigenous peoples and *mestizos* appear to the traveler or migrant as intriguing, even perplexing, but usually untrustworthy. (4) Additionally, the protagonists have to suffer a "savage" extractivist capitalism, mafioso and even genocidal, which has preceded them and constantly threatens them. From rubber to logging, oil to natural gas, gold to coca leaves and soybean plantations, the extractivist industries threaten to enslave travelers and migrants alike, just as they have done with native peoples. (5) The national state has a very limited presence in the region for enforcing the rule of law and is usually caught up in corruption.

Thus, due to its socio-environmental impacts on the indigenous and nonindigenous peoples, the *novela de la selva* is a genre that criticizes the expansion of extractivist industries into the Amazon as well as the role of colonial and neo-colonial states in such expansion. In spite of this indictment, Amazonian indigenous peoples are never protagonists in this genre. In fact, they always incarnate the various definitions of the "primitive" in vogue in the twentieth century. In spite of claiming to reveal the truth of what is going on in the Amazon, the *novelas de la selva* are always making the case that all sorts of travelers and migrants failed because of a lack of critical knowledge of the place: of its history, its peoples, and its ecologies. Humboldt qualified the tropical rainforest as Impenetrable. A hundred and fifty years later, Carpentier called the jungles of the Orinoco River the Unknown. Now socio-environmental scientists and humanists are working on scenarios and procedures for dealing with its Uncertainty. Those alternative modernities pursued by the protagonists of this genre clashed with "unnatural" forms of sociability and communication among humans and nonhumans. These interactions are difficult to accept, philosophically speaking, for a modern subject but seem to be possible on the ground. There is, however, a more horrifying lesson emerging from these narratives. The crisis or failure of these endeavors carried out in the Amazon also means the collapse of what each alternative project understood as the appropriate, or the "this-time-we-got-it-right", relationship between nature and some version of what modernity ought to be for producing wealth and/or well-being. It is just an error, but an error for which you pay with your own life or with the lives of others.

Read in this way, the *novela de la selva* anticipated fundamental assumptions in current environmental studies. Environmental crises have led us to the awareness that, as humans, we are inserted in communities with other species and beings that emerge from our entanglement with these agential beings. The adjective "socio-ecological" does not name two separate domains that occasionally interact. It is a pleonasm that wants to stress the imperative need to develop knowledge that integrates these domains for both the understanding and the solution of environmental problems. Additionally, if some readers find this environmentalist reading of the *novela de la selva* to be a useful method to reflect upon the complex impact of economic development policies, as well as on the complexities stalking what looked like straightforward alternatives to such policies, then this is a good time to introduce the project of the Environmental Humanities.

The Amazonian Turn

The term Environmental Humanities (EH) is still new outside the Anglophone world. The rise of the EH is linked to an increasing consensus in sustainability and resilience studies that suggest the following: scientific and economic arguments are not having the expected impact in public debates, and the narratives that have emerged, especially apocalyptic ones, fall short for fostering change or may even prove counterproductive. New narratives are needed, perhaps even ones that embrace ambiguity. Secondly, the consensus points out that the understanding of the weight of beliefs, moral values, ethical dilemmas, emotions, aesthetics, and creativity in environmental conflicts is not sufficiently sophisticated among analysts, politicians and the public. Environmental humanists are calling for "a deeper understanding of how less tangible, nonmaterial values shape management and stakeholder decisions" (Kitch, p. 2).[1]

The arrival of the Environmental Humanities is also helping to validate, give visibility, and facilitate the convergence of the many approaches already out there focusing on the power of culture and the arts as tools for transformation, and which often predate the rise of the term itself. There have been decades of research on either the cultural roots of problematic uses of the environment, or alternative ways of conceiving the association and interaction between humans and nonhumans, in the history or present of different peoples (modern or nonmodern). Regarding the role of human and nonhuman identities, responsibilities, and affects as part of the conflicts as well as of their solutions, the questions that drive the EH can also be found nowadays across specializations, disciplines, activisms, cultures, and politics. These less tangible, nonmaterial values are also a priority in the environmental agenda of social movements. If it is the case that similar concerns that we usually identify with the humanities are taking place in parallel scenarios, then it is reasonable to conclude that the EH is part of a larger phenomenon worthy of attention and reflection. As such, the EH should keep a critical eye on itself. At the same time, even if we are witnessing competing discourses, the point now seems to be to find ways and join scenarios in which the EH could engage with such plurality.

Let us turn now to the "Amazonian turn". By it, I understand, first, the impact of Amazonian cosmologies of the human and the nonhuman on the "new" political ontologies attempting to redefine what we call "natural" and "social" and the questioning of the universality of the underlying notions of object and subject. Amazonian anthropology has helped us understand that we can no longer assume that, as French anthropologist Philippe Descola says,

> Our own reality, our ways of establishing discontinuities and detecting stable relationships in the world, our manners of distributing entities and phenomena, processes and modes of action into categories which are allegedly predetermined by the texture and structure of things, are a universal fact of human experience. (pp. 30–31)[2]

Since the foundational work of Brazilian anthropologist Eduardo Viveiros de Castro on Amazonian perspectivism and Descola's work on Amazonian animism,

---

1 Kitch, Sally L.: "How Can Humanities Interventions Promote Progress in the Environmental Sciences?". *Humanities* 6(76), 2017, pp. 1–15.
2 Descola, Philippe: *The Ecology of Others*. Prickly Paradigm Press: Chicago 2013.

Amazonian cosmologies or ontologies have been informing the efforts of the so-called new materialisms to create metaphysical alternatives to the privileging of the modern dichotomy between nature/society as a theoretical tool, and to avoid the naturalist or culturalist reductionism of human/nonhuman relations that this dichotomy supports. The "Amazonian turn", therefore, is part of a larger effort to redefine the object of study of the natural and social sciences and of the humanities as well (not only of the environmental humanities). Let us take note, for the time being, that these efforts, like in Bruno Latour's *An Inquiry into Modes of Existence: An Anthropology of the Moderns* (pp. 233–257), also include a revision of the ontological status of fictional, mythological, and imaginary beings in general, and therefore, of the notion of text in modern and postmodern literary theories.[3] Instead of emphasizing that these beings are "mental" entities supported by language, the "new ontologies" invite us to think about the experience of these beings akin, for instance, to the perception of invisible human and nonhuman beings in Amazonian shamanism. In any case, the point that I want to stress now is that the "Amazonian turn" is at least an invitation to discuss the redefinition of the object of study of Amazonian natural sciences, social sciences, and humanities. Indeed, for starters, it is helping us to understand certain values, beliefs and aesthetics driving current popular environmentalisms in the region.

This recent research in political philosophy shares some basic positions with the agenda of numerous indigenous movements. Both argue for the status of the nonhuman as a subject or at least an "actant", rather than an object, of politics. Both reject the nature/culture dichotomy. And both critique the naturalist or culturalist reductionisms in which this dichotomy eventually occurs. Although it is wise to be always alert against appropriations of indigenous cultures, this concurrence of opinions helps to explain that the bold claims made by indigenous peoples on the ontological status of the nonhuman are not coming from a naïve primitivism or an outmoded romanticism. Rather, they derive from the recovery of beliefs, values, rituals and other practices that indigenous communities and organizations, like many other people on the planet, have concluded is the best way to deal with the negative socio-eco-cultural impacts of the economy of development. It is not simplistic to say that indigenous politics have their own sort of EH agenda. Traditional values and beliefs are not there merely for stopping or slowing down the intervention of capital and opposing the state's

---

3   Latour, Bruno: *An Inquiry into Modes of Existence: An Anthropology of the Moderns.* Harvard University Press: Cambridge, MA 2013.

complicity in the name of indigenous identity. They are invoked fundamentally as a path for fostering well-being and happiness.

This "Amazonian Turn", reflecting the role of Amazonian cosmologies in academic debates on political ontologies and the current resistance by indigenous peoples and small farmers on the ground to this extractivism, suggests new leads for further understanding the *novela de la selva*, Amazonian literatures, and arts in general. "Amazonian", for the purposes of this chapter, is understood very broadly: indigenous or nonindigenous written narratives, written by scientific travelers, state officials, or intellectuals; or indigenous and *mestizo* oral narratives, transcribed in many different fashions by anthropologists, writers, educators, and the indigenous peoples themselves. In sum, we suspend for a moment the distinction between literature written by "insiders" and "outsiders". At the convergence of the impact of popular Amazonian environmentalism in national politics, and of the philosophical debates on the political ontologies implied in those social movements, emerges the need to ask now if the notions of human and nonhuman, and of text as a cultural construct, that so far have often supported ecocritical readings, including of Amazonian literatures and film, are actually in line with Amazonian ontologies. Is it possible to propose a theory of Amazonian textuality that would not rely on a notion of text built upon the split between nature and society, human and nonhuman?

Amazonian Textuality

Any theory of Amazonian literatures and arts cannot ignore the lead of Amazonian animism of the nonhuman and will have to follow it beyond indigenous texts, art, and film. It will require doing more than just finding the actancy of nonhuman beings in the narratives' plots. It forces us to revise our own assumptions on what we, in the humanities, consider the object of our studies. Any definition of cultural entities and events in the humanities is built upon its ontological opposition to the modern notion of nature, even if we consider "culture" to be an interface between society and nature, because culture belongs to the domain of the subject as it is defined in the dichotomy between subject and object. We are learning how to speak about objects as agents, actors, actants, interveners, etc., however, in ecocritical studies in Spanish, it seems to me, we are leaving the subject untouched. It is still right to say, in postmodern and poststructuralist fashion, that the genre of the *novela de la selva*, under the right analysis, reveals its own self-contradictions. For instance, at the beginning of this chapter, I mentioned its failure to reach and secure alternative forms of embracing modernity in close contact with the landscape. It is still right to say, also in postmodern and poststructuralist fashion, as Carlos J. Alonso did a long time ago in *The Spanish American Regional Novel: Modernity and Autochthony*, that the text is not an "autochthonous" being rising from the landscape, but that such assessment is the result of an accompanying and authoritative discourse of nature; therefore, there is no "organic" or "natural" link of the written text with the environment.[4] Nevertheless, in the current state of affairs of popular environmentalisms, and the conversation about more political ontologies for facing collectively and interculturally the Anthropocene, we cannot allow such readings to become our only paradigm for thinking about the *novela de la selva* or any Amazonian cultural entity, indigenous or not. The readings above ontologically split the text from the forest, by making the forest something physical and biological, and the text, an event of symbolic communication that originates only in human subjects. As such, the forest to which the text refers is supposed to be a "being of fiction" that only exists in our minds constructed by language.

Is it possible to have an Amazonian ecocritical theory that would not rely only on a modern or postmodern notion of text built upon the split between nature and culture? Amazonian anthropology has been working on this. Eduardo Kohn, in *How Forests Think: Toward an Anthropology beyond the Human* resorts to ecological phenomenology to translate the fact that, among the Runa people of Ecuador, every living being in the forest is able to communicate but not necessarily in human symbolic systems.[5] In this semiotics of the forest, every being is capable of sending a message, and in fact they do so, although not necessarily with human words. It could be done through what the American philosopher Charles Peirce called indexes and signals, such as smell, taste, touch, emotions, and dreams. Or, nonhuman-to-human communication can also happen through music, dance, singing, designs, drawings, rituals, etc. These messages can be converted into words, including alphabetical writing. No medium is deliberately excluded. On the contrary, Amazonian textuality welcomes resorting to multiple media and modes of communication.

Not only the field of Amazonian anthropology, but the environmental humanities in general seem to be going in the direction of a more comprehensive model of more-than-human semiosis, that is to say, to not necessarily privilege theories of texts and cultural representation following the model of a text as a human subject's enunciation, or as a human speech act in a symbolic human language.

---

4 Alonso, Carlos J.: *The Spanish American Regional Novel: Modernity and Autochthony*. Cambridge University Press: Cambridge 1990.
5 Kohn, Eduardo: *How Forests Think: Toward an Anthropology beyond the Human*. University of California Press: Berkeley 2013.

Symbolic languages image worlds by virtue of the ways in which the sign is embedded in a symbolic system, and that system constitutes the absent context for the meaning of any given word's utterance (Kohn, p. 37). We can instead bring into our conversation about cultural beings the idea of the icon: where the "sign vehicle" resembles, for the interlocutors, the entity it represents not out of naïveté, but because the interlocutors are willing to ignore the difference, and where the icons are what they are 'in themselves' regardless of the existence of the entity they represent.

Michael Uzendoski, another anthropologist who has also studied among the Runa, in "Beyond Orality: Textuality, territory, and ontology among Amazonian Peoples", insists that, in the Amazonian world, the text is the dialogical and intersubjective presence of the landscape and that this presence is recreated or activated experientially by storytellers when they tell myths, recount experiences, describe paths, or explain the action of people.[6] The text is the cointeraction of human, plant, animal, and spiritual agential beings. Without the agency of these diverse territorial subjectivities and their influence upon the human subconscious, Amazonian texts would not cohere as meaningful works of art and experience. There is nothing preventing "outsiders" from experiencing such communication and interactions. The text does not originate in a human language, nor is its message the medium that co-interacts with the text. The cosmic experience that the text facilitates is trans-linguistic. In Amazonian storytelling, argues Uzendoski in "Fractal Subjectivities: An Amazonian-Inspired Critique of Globalization Theory", the fundamental rule is the embeddedness of the subject in large cosmic flows of exchange relationships between the human domain, the animal, the landscape, and the spirit world.[7] Humans and nonhumans are not qualitatively different beings. They all possess a soul, intelligence, and the qualities that we attribute to subjectivity (animism). Nonhumans see other members of their same group as humans, whereas members of other groups are seen either as prey or predator (perspectivism). Animals and spirits are unpredictable and dangerous because humans do not really know what animals and spirits want or when they may strike (dark shamanism). "Contemporary Amazonian storytellers—explains Uzendoski—conceptualize themselves as fully modern

---

6 Uzendoski, Michael: "Beyond Orality: Textuality, territory, and ontology among Amazonian Peoples". *HAU: Journal of Ethnographic Theory* 2(1), 2012, pp. 55–80.
7 Uzendoski, Michael: "Fractal Subjectivities: An Amazonian-Inspired Critique of Globalization Theory". In: Hutchins, Frank / Wilson, Patrick C. (eds.); *Editing Eden: A Reconsideration of Identity, Politics, and Place in Amazonia*. University of Nebraska Press: Nebraska 2010, pp. 38–69.

subjects, but they define their subjectivity through fractal relationships with animals, spirits, and nature" (2010, p. 39). A fractal is a particular, well-defined, and easily repeated set of rules occurring in infinitely receding or expanding scales. The rules of the cosmic body, animism, perspectivism and dark shamanism appear in stories about local Amazonian native and nonnative lives and in Amazonian stories about national identity or globalization; "These relationships—adds Uzendoski—do not oppose locality and globality, they show that people are defined simultaneously by the local and the global, part and whole, and the one and the many" (2010, p.39).

In Amazonia, native and nonnative peoples have shared space, intermarried, and told stories to one another at least since the beginning of the Rubber Boom in the late 1880s (Uzendoski, 2010, p. 55), precisely at the time of the emergence of the *novela de la selva*. The corpus of Amazonian storytelling is so vast and recurrent in diverse locations that their linear connections (their origin, whether native or nonnative) are in chaos. "But the order appears at the level of underlying relations" (Uzendoski, 2010, p. 58). This expansive and border-crossing quality of Amazonian storytelling and arts led Uzendoski to propose the notion of "Amazonia at large". What I am proposing is that the texts and artifacts in Amazonian literatures and arts could be read as simultaneously Western and modern, on the one hand, and in an Amazonian relationship with animals, plants, and forest spirits, on the other. Precisely, contemporary Amazonian life invites us to cross boundaries between natives and nonnatives, the many ethnicities, the nation-states boundaries, indigenous and nonindigenous languages, academic disciplines, literary genres, media, etc. And, above all, the human and the nonhuman. Throughout Amazonia, one can find a continuum of native-nonnative relationships and subjects that are rich and complex (Uzendoski, 2010, p. 55). For instance, the *ribereños*, or river people, although with an indigenous cultural lineage, no longer possess an overt "native" identity or speak a native language; however, they do have a deep knowledge of Amazonian ecology and cultural patterns. Or let us consider the popularity of *mestizo* urban shamans in the fast-growing cities of the Amazon. It speaks of the adaptation of nonnatives to native concepts regarding the relationship with nature. Or the case of indigenous and *mestizo* shamans who often teach outsiders to interpret their experience from a very specific Amazonian perspective. It is not unusual for native and mestizo shamans to grant "cosmic" belonging to the Amazon to visitors, travelers, and anthropologists who come to spend some time with their communities (Uzendoski, 2010, p. 57).

In spite of its poor or null representation of indigenous voices and epistemologies, the *novela de la selva* stages, through nonnative characters, the principles of the cosmic body, animism, perspectivism, and dark shamanism. There is an immersion experience behind these texts, usually a very extensive one, which most likely had consisted of telling and retelling stories among nonnatives and natives. But these texts were also the result of an extensive encounter with non-anthropic beings. Along with Eduardo Kohn, I would say that for writers and characters in the *novela de la selva*, the phenomena in the forest become, often reluctantly, signs that stand for "someone". The texts are listening to what Kohn calls the Amazon *sylvan selves*. Certain perceptions, dreams and visions, or irrational behaviors in the characters that look, to the conventional literary critic, as remains of a by-gone Romanticism, or the result of pathological epistemologies (characters usually get sick with fevers), are in fact experiences of animism, perspectivism, dark shamanism, and the cosmic body exchange. In the town, the mission, the river, and of course the forest itself, the white characters react verbally and psychologically to the overwhelming physical-biological-social environment that defeats their attempt to control it. Subjects who believe in the modern difference between nature and society, and in fact preach a life free from the forces of nature, are forced into the troubling "subjectivization" of the nonhuman. The inability to name this excess to meaning leads to rage against the jungle and name calling of these sylvan selves ("Green Hell", for instance). Nevertheless, the whole process undoubtedly points to the fact that nonhuman actors are modifying other actors, including actors endowed with consciousness, speech, will, and intention. The *novela de la selva* is not asserting Amazonian ontologies by translating and retelling them, but by becoming subjected to them. Of the time period when the *novela de la selva* arose, we might say similarities apply to the current state of affairs in the Amazon and "Amazonian realities are present and active in defining the subjectivity of the nonindigenous who now represent the majority of the population in many regions" (Uzendoski, 2010, p. 59).

Embracing the Serpent

Neither the *novela de la selva* nor Amazonian literatures and arts in general need to be theorized or historicized with only the categories and frameworks typically found in academic literary studies. Furthermore, such confluence of perspectives is often staged in the fictions themselves. The film *El abrazo de la serpiente* (2015) [The *Embrace of the Serpent*] by Colombian director Ciro Guerra is an exemplary case. Although the film is obviously not a *novela de la selva*, its plot is set in the historical period of the Rubber Boom and follows many features of the *novela de la selva* genre, while contesting others, but above all it does a better job than those narratives in (a) showing how Amazonian human and nonhuman realities were defining nonindigenous subjectivities at the time, and (b) showing how the same cultural texts can be "read" under a semiotic framework that includes, but does not privilege, the symbolic.

The film has accumulated 29 wins and 14 nominations in film festivals throughout the world, including the 2016 nomination for an Academy Award in the Best Foreign Language Film category. Judges and reviewers celebrated Guerra's film because in it, against the filmic and literary tradition, the protagonist is not a Westerner, and the story is told from the viewpoint of the indigenous characters who speak in their native languages. Additionally, the critics and the film promo praised the screenwriters because, after three years of research, the script was developed in consultation with native tribes. Furthermore, while shooting the film, Guerra himself was open to suggestions made by the indigenous actors that play some of the main roles: Nilbio Torres, of the Cubeo ethnic group, who plays the young shaman Karamakate; Antonio Bolívar, one of the last survivors of the Uitoto ethnic group, who plays the old Karamakate; and Yaukenü Migue (or Miguel Ramos) who plays Manduca, the guide and assistant to the German anthropologist Theodor Von Martius. Clearly, the purpose of these side narratives is to establish that the film tell a story about colonialism from the perspective of one subject who has radically resisted it. Additionally they also state that the film itself is not the outcome of a colonial relationship between producers, director, and scriptwriters, on the one hand, and the indigenous actors and communities who supported the production of the film, on the other.

Shot in black and white, Guerra's film is inspired by the diaries of two botanists: the German Theodor Koch-Grünberg (1872–1924) and the American Richard Evans Schultes (1915 -2001). The film's main protagonist is Karamakate, an indigenous shaman, the last survivor to the fictional tribe of the Cohiuano, which has been devastated by the exploitation of the so-called barons of the rubber industry in the twentieth century (1879–1912, and then 1942–1945). The film alternates between the two stories. The first one is the journey of a young Karamakate with Theodor von Martius (the character partially inspired on Koch-Grünberg) and Manduca, his indigenous assistant. Theo is a sick German anthropologist seeking the help of Karamakate's healing powers. His last hope is to be cured with *yakruna* (a hallucinogenic plant), and that Karamakate would lead him to its secret location. Theo has risked his life in the Amazon pursuing a dream of a mysterious geometrical figure that he first had back in Germany, in fact the only dream that he has ever had, even under the influence of hallucinogenic plants.

Karamakate, though, refuses to help. He has lost his tribe, and with that, any sense of purpose other than utterly rejecting any contact with the whites, treating them with a contempt only comparable to the self-righteous attitude he reserves for other indigenous peoples who collaborate with white characters. Karamakate despises Manduca because the latter's tribe has "submitted to the whites without a fight". Manduca, however, argues that Theo is an ally to the survival of indigenous peoples, and supports his work in the hope that it would help in advocating for their rights against the discriminatory policies of the nation-state. In spite of Karamakate's scorn for both, Theo finds a way to recruit him by offering a deal he cannot refuse. In exchange for yakruna and Karamakate's healing powers, Theo will take him to the place where the last Cohiuanos live. After many adventures, upon the arrival to their final destination, Karamakate, enraged at the environmental and moral degradation in which the last Cohiuanos reside, destroys the last remaining yakruna plants that the Cohiuanos are cultivating now as a recreational drug. Suddenly, the village is attacked by an army working for the rubber factories. In the chaos that ensues, Karamakate ignores Theo, who at this point has to be carried by Manduca. Thanks to him, Theo escapes from the village and both return to the river only to eventually die, as we learn through the alternate story.

Forty years after, an old Karamakate embarks on a new journey, this time with Evan (the character inspired in Schultes), an ethnobotanist who is in search of the yakruna to use it to improve the production of rubber during World War II (Evan hides this motivation from Karamakate). As a matter of fact, Evan is after the yakruna in the hope that the plant would facilitate an experience that he has never had. Evan claims that he has never dreamed, not even with the help of other shamans and hallucinogenic drugs. Karamakate agrees to help Evan because it is the American ethnobotanist who actually would be able to read the signs that show the way to where the last yakruna plant is located. The sacred glyphs in the rocks, the plants, and the animals no longer speak to the old shaman, and because of this silence, old Karamakate concludes that he has become a *chullachaqui*, that is to say, just an image, the appearance of a person, empty and destitute of any memories. Upbeat because of the chance to recover his memory, but also because of the opportunity to fix the mistake made forty years earlier with another scientist, Karamakate allows Evan to guide him. For Karamakate, Theo and Evan are the same subject. Both travelers are connected at the highest levels of Amazonian reality. And finally, at that same level, old Karamakate finally understands that his particular mission was, in spite of his rage against colonialism, to share his knowledge with travelers like Theo and Evan.

Embrace of the Serpent carries a story in which an Amazonian *thinking forest* clearly intervenes in the way indigenous and nonindigenous characters conceive of their identity and the way they understand the meaning of their lives. Paradoxically, it is easier to explain such intervention when it comes down to the nonindigenous scientists than to Karamakate himself. For neither Theo nor Evan are Amazonian cosmologies merely somebody else's beliefs by which those who believe in them think they are able to act upon the world. Even if only out of personal desperation, Amazonian cosmologies have become, for them, legitimate knowledge and alternative and complete systems of conceptualization of the world to act upon. Although Karamakate is of course more familiar with Amazonian cosmologies, it turns out that it is more difficult for him to interpret the messages sent by the forest. Karamakate's remarks about the beliefs and behaviors of the scientists, and whites in general, intend to castigate their cosmology and to ascribe to it all the evils with which modern subjects are afflicted. There is in Karamakate an ethnocentrism and rejection of the European other explicable by the history of colonization, but in conflict with what apparently the forest wants him to do. This reverse ethnocentrism is not the only issue to bring forward for discussion. In a sense, the film takes us back in the history of representation of indigenous peoples, particularly of Amazonian peoples, to an essentialist and maybe nostalgic representation of a beautiful naked and fierce Indian who rejects the outsiders because of their association with colonialism. The audience must be warned against the fact that such an essentialist representation of an ideal indigenous identity and resistance is in conflict with current indigenous peoples' social movements in Amazonia and elsewhere. Instead of retreating to isolation, indigenous peoples are defending their territorial rights and ways of living not only in the forest but in the courts, the parliaments, and the media. Consequently, for their effort, they are proving to be able to mobilize nonindigenous sectors within the nation-state and abroad.

In spite of the viewer's fascination with young Karamate, I do not believe that the film is, after all, a celebration of Karamakate's justified rage against colonialism. The film's plot is a story that illustrates the tension of an intercultural experience regarding the overcoming, but also the maintaining, of difference. It is a quest narrative of two (or perhaps four) subjects, indigenous and Western-modern, who happen to be cornered into extreme situations of self-alienation by the consequences of several waves of colonialism in the Amazon. Driven by the desire to discover or recuperate the hallucinogenic plant of the *yakruna*, and to experience dreaming or remembering, according to each case, these subjects join forces and make a fragile negotiation that is constantly challenged and reestablished according to the circumstances on the road. Indeed, the negotiation does not prevent them from criticizing their respective ontologies, moral values, and ethical behaviors. Nevertheless, both indigenous and nonindigenous characters are subjected to the agentic powers of the forest itself, which communicates with them in a variety of ways, but whose messages are difficult to interpret. Or, surprisingly, they are messages that Karamakate has deliberately denied. Thanks to a process that requires the intervention of outsiders, old Karamakate finally realizes that his mission was not just to pass on his knowledge to his own people but also to impart to others a deeper understanding of the forest and to awaken a deeper spirituality in them. He finally lives up to the expectation that many years earlier Karamakate himself demanded from Theo: "Knowledge belongs to all. You do not understand that. You are just a white man."

There is a scene in the movie remarkably useful for illustrating the conflict, but also the coexistence, between Amazonian and non-Amazonian ways of creating meaning for a cultural artifact. Young shaman Karamakate sees a black-and-white photograph of himself taken by Theo von Martius. After taking a look at it, Karamakate does not want to return the picture to Theo. Theo explains, to Karamakate's surprise, that the picture belongs to him. "But it's me", replies Karamakate. "It's not you. It's an image of you", Theo responds. Karamakate reaches the conclusion that the photograph is a *chullachaqui:* "We all have one. He looks just like you, but he's empty, hollow". Theo defends the picture as something more than an empty sign: "This is a memory. A moment that passed". Still, Karamakate resists Theo's explanation of the picture: "A *chullachaqui* has no memories. It only drifts around in the world, empty, like a ghost, lost in time without time". Many years later, when encountering Evan, old Karamakate wonders if he has not become a *chullachaqui* of himself, because of his lack of memory. Confident that the picture is just a *chullachaqui*, Karamakate allows Theo to take it back with the purpose of showing it to his people in Germany. This use of the term *chullachaqui* is unusual. Although in Amazonian cosmologies the *chullachaqui* is a complex guardian of the forest, in his definition of the term Karamakate resorts only to one of its characteristics: the *chullachaqui* disguises himself under the appearance of somebody close to the person who has gone alone too deep into the forest. In any event, placing this comment by Karamakate about a black-and-white photograph of himself taken by a white foreigner into a black-and-white cinematography of another white foreigner who deliberately evokes the pictures taken by Theo, raises the question if, from an Amazonian ontological perspective, this very film could be a source of memory, truth and life.

Karamakate's initial interpretation of the photograph is iconic in the sense in which the term is used by anthropologist Eduardo Kohn for studying the semiosis of the Amazonian forest. The picture is Karamakate because of the virtue of the resemblances between the "sign vehicle" (the photograph) and the object (himself), and because Karamakate is willing to ignore the differences (Kohn, p. 31). The issue here is not the inability of this young and self-isolated shaman to understand a symbolic system of communication. What Karamakate's response is doing is to give visibility to the iconic and other nonverbal and nonsymbolic modes of communication and representation. Both young and old Karamakate criticize and even show their contempt for the media and archives carried by the scientists with great effort, up and down the rivers, as obstacles to achieving true wisdom because these tools of symbolic languages separate the travelers from the world surrounding them. The film itself, then, suggests another framework for considering its relation to reality and its experiential force. Point well taken, but the point is not to privilege the Amazonian icon over other forms of semiosis. Karamakate's picture becomes more than a *chullachaqui*, or more than one of those colonial photographs famous for "fixing 'native' peoples into images, and unilateral narratives and depictions" (cf. Fernandez de Lara Harada) because this film offers a rich and complex narrative that the photograph of Karamakate would never be able to carry. If this is the case, then, the movie would vindicate as indispensable all the symbolic tools thanks to which there is a narrative into which we can contextualize the photographs.

Conclusion

Do we really need to push this hard regarding the ontology of the Amazonian text? Is it not enough to be able to read non-Modernist ontologies of the human and nonhuman within what the text narrates? Let me go back to the beginning of this chapter. Why have these narratives on extractivism become so irrelevant outside literary studies (or even within) despite some of them being very famous? And let us narrow it down to the case of Peruvian Amazonian literature. There are no Amazonian texts in the Peruvian literary canon. It is well known that the *novela de la selva* was scorned by Vargas Llosa and Carlos Fuentes in the 1970s for allegedly being more geography than literature interested in the affairs of the modern world, which had become essentially urban. However, this exclusion of Amazonian literature from national literary canons has also happened in spite of widely accepted theories on Latin American literature that have fostered the development of inclusion in so many instances (with notions such as Ángel Rama's "transculturación narrativa" and Antonio

Cornejo Polar's "literaturas heterogéneas"). Yet, Amazonian literatures and arts, indigenous or not, remain in the margins. In any event, paradoxically, in the field of history of Latin American literature, the more it has wanted to be inclusive and successfully done so, the more it has contributed to the invisibility of Amazonian literatures. Why? Because literary studies have neither been able nor willing to appropriate them for imagining alternative national identities, or for asserting alternative forms of modernity, as it has been done, for instance, with the Andean world. And because these theories were looking for those meanings in the author's use of symbolic languages: how European and indigenous languages were brought together, how dialects played together within the text, and how the structure of the narrative discourse responded to the input of conflicting cultural traditions. It seems to me that overall Amazonian literatures in Andean countries were less interested in becoming a sort of national aesthetic avant-garde. An Amazonian ecocritical theory, based on alternative notions of textuality, would help us explain and overcome the systematic absence of Amazonian texts in every national literary canon in the Andean/Amazonian countries. Literary studies, even when taking a critical position regarding the role of culture under neo-liberalism, has been part of a political regime based on the nature-society dichotomy, and has actively resisted, as a sign of pre-modern atavism, "earthing" or embedding identity and texts and cultural entities in environments. Maybe literary studies have been an unwilling accomplice in a war against beliefs and practices that ignore the separation of entities into nature and culture. Thus, an Amazonian ecocritical theory would also mean involving the Humanities in the larger political experiment of building a common world in countries characterized more than by their cultural plurality by the plurality, but of ontological worlds coexisting or in conflict.

Bibliography

Alonso, Carlos J.: *The Spanish American Regional Novel: Modernity and Autochthony*. Cambridge University Press: Cambridge 1990.

Descola, Philippe: *The Ecology of Others*. Prickly Paradigm Press: Chicago 2013.

Fernandez de Lara Harada, Jessica: "Embrace of the Serpent". Retrieved 6.20.2016, from Gates Cambridge website.

Guerra, Ciro (dir). *El abrazo de la serpiente*. 2015. Buffalo Films/Caracol Televisión.

Kitch, Sally L.: "How Can Humanities Interventions Promote Progress in the Environmental Sciences?". *Humanities*, 6(76), 2017, pp. 1–15.

Kohn, Eduardo: *How Forest Think: Toward an Anthropology beyond the Human.* University of California Press: Berkeley 2013.

Latour, Bruno: *An Inquiry into Modes of Existence: An Anthropology of the Moderns.* Harvard University Press: Cambridge, MA 2013.

Uzendoski, Michael: "Beyond Orality: Textuality, territory, and ontology among Amazonian Peoples". *HAU: Journal of Ethnographic Theory* 2(1), 2012, pp. 55–80.

Uzendoski, Michael: "Fractal Subjectivities: An Amazonian-Inspired Critique of Globalization Theory". In: Hutchins, Frank / Wilson, Patrick C. (eds.): *Editing Eden: A Reconsideration of Identity, Politics, and Place in Amazonia.* University of Nebraska Press, 2010, pp. 38–69.

Laura Barbas-Rhoden

Gendering Ecohispanisms: Knowledge, Gender, and Place in a Pluricultural Latin America

Abstract: This chapter builds upon robust research in the fields of gender studies and ecocriticism by exploring intersectional ecocritical studies as one of the most relevant areas for inquiry into Latin American cultural texts.[1] Broadly speaking, the study of the myriad ways that gender, environment, power, and knowledge interrelate and mutually co-constitute one another in the pluricultural world of Latin America is a decolonial and depatriarchalizing praxis with profound implications for ecocriticism. In highlighting this kind of ecocritical practice here, I advocate for the broadest possible understanding of gender identities, as well as an expansion of the definition of environment, to refer to territories, places, landscapes, and most generally, the material world in which human communities and individual human beings live their lives. This broadening of definitions serves the purpose of repositioning Western European and Anglo-American ways of thinking about being and knowing in a pluriversal cultural universe, and it insists that gender matters as a category for consideration with regard to all cultural texts (not only those authored by women or labeled as ecofeminist).

Keywords: Gender Studies, Ecocriticism, Latin American Culture, Environment, Materialism

Although to date the most generally recognized field of literary inquiry related to gender and environment has been ecofeminism and, more recently, feminist ecocriticism (Gaard / Estok / Opperman 2013), interpretive questions at the intersection of gender studies and ecocritical studies must be asked of more texts, from and about more places, and in more languages.[2] In fact, the continuous decolonization and depatriarchalization of ecocriticism and Latin American

---

1 Kimberlé Crenshaw coined the term "intersectionality" in 1989 to explain the multiple forms of power and exclusion that shape the experiences of African-American women in the United States. The term has since come to be widely used in the social sciences and humanities in reference to the way power operates in complex ways. Crenshaw, Kimberle: "Demarginalizing the Intersection of Race and Sex: A Black Feminist Critique of Antidiscrimination Doctrine, Feminist Theory and Antiracist Politics". *The University of Chicago Legal Forum* 140, 1989, pp. 139–167.
2 Oppermann, Serpil. "Feminist Ecocriticism: A Posthumanist Direction in Ecocritical Trajectory." In: Gaard, Greta / Estok, Simon C. / Opperman, Serpil (eds.): *International Perspectives in Feminist Ecocriticism*. Routledge, New York 2013, pp. 19–36. Opperman

literary and cultural studies rests on the pursuit of questions such as these: What knowledges, rhetorics, and discourses about the body politic, body human, and bio-geo-physical bodies appear, morph, or disappear across the continuums of space and time in always already gendered imaginations from and about the Americas? Which are privileged, which marginalized, which suppressed, recovered, or repurposed, in what ways, and to what ends? What differentials of power mediate such dynamics in the cultural sphere, and what are the material consequences of these dynamics in specific historical moments? Effectively, the pluricultural, pluriversal reality of Latin America invites the interrogation of gender, environment (or nature), knowledge, and culture as mutually constituted categories, and products from this inquiry will challenge and enrich ecocritical practice everywhere.

Substantial scholarship in gender studies has led to an understanding in the twenty-first century academy that constructions of masculinities, femininities, and indeed, a full spectrum of identities and expressions of gender are contingent and dynamic, intersecting importantly with constructions of ethnicity, race, class, geopolitical location, religion, and age. Latin Americanists like María Lugones (2010) have underscored the important ways in which coloniality has shaped understandings of each of these categories.[3] Expanding concurrently with the increase in scholarship in gender studies is the corpus of theoretical and critical perspectives about the ways diverse human communities, contemporary and historical, understand human relationships in relation to the material environment of which they are a part (Escobar 2008; Escobar 2015; Gaard, Estok, and Opperman 2013). Existing and future ecocritical work of an intersectional nature, in the context of Latin American literature and cultural studies, is therefore situated in a space in which multiple fields of disciplinary inquiry overlap and in theoretical and institutional contexts in which disciplinary boundaries increasingly are blurred and challenged, as in the case of decolonial studies (Mignolo 2009; Mignolo 2011a; Mignolo 2011b). Intersectional ecocriticism thus exists in a space of complexity and dynamism, propitious for a pluralism of approaches.

Intersectional ecocriticism in the Latin American context is premised upon the understanding that the power dynamics of gender, coloniality, and geopolitics (among others) shape not just the material world, but also what is acknowledged

---

offers here a cogent explanation of the lineage and scope of feminist ecocritical approaches.
3   Lugones, María: "Toward a Decolonial Feminism". *Hypatia* 25(4), 2010, pp. 742–759.

as legitimate knowledge about the material world; what are legitimate processes by which knowledge is produced; and who (or what beings) may know and act. In Latin America, as in other areas, particularly those with a colonial past, hierarchies of ethnicity, class, race, and gender organize and control landscapes, and knowledge of landscapes and territories have been constituted by means of a dynamics of power inflected by the interplay of ethnicity, class, race, and gender. Furthermore, notions of both place and identity (identity in which gender and also place, or lack-of-place, have a role) have been organized, legitimized, and delegitimized according to the values, norms, and assumptions of dominant groups, both in the macro realm (say, of "the academy" worldwide) and within each particular cultural and social group, historically and in the present. As ecocritics acknowledge that knowledge, gender, territory or place, and power are co-constituted in complex ways and develop frameworks and language by which to express that acknowledgement, the field itself is reshaped to be more inclusive of cultural texts and critical perspectives from various historical periods and places.

Though intersectional ecocriticism may take various directions, it can generally be expected to frame analysis in terms of emplacement, contingency, and networks, in which gender identities, roles, and norms are mutually co-constructed with other aspects of self and society in material and immaterial spaces; to acknowledge dynamism and interplay of diverse forces that give rise to what humans believe themselves and others to be; and to emphasize that forces operating at different levels of scale (local, regional, national, transnational, to use contemporary political labels) shape knowledge, being, and action. In considering what has been counted, in different times and places, as "legitimate" knowledge, or experience, or living, intersectional ecocriticism will continue to explore the philosophical substrates (epistemological, ontological, ethical, phenomenological) that inform texts, traditions, and canons and to tease out how texts represent being, knowing, doing, and/or discerning by diverse human actors and by other-than-human or more-than-human agents. It can undertake close readings of representations of a given phenomenon (volcanic activity in El Salvador; the representation of the caudillo and land in novels; water in Amazonia) within a given corpus of diversely authored contemporaneous texts, or it might consider the change in representation of a phenomenon over time by authors that share certain aspects of identity or experiences.

Reading Latin American cultural texts ecocritically, with an eye for gender as a category of analysis, implies a constant grappling with (1) cultural heterogeneity, and the epistemological heterogeneity that is its substrate, (2) multiple levels of scale, from the local to the regional, national, and global, and (3) the dynamism of sociocultural realities, as these are shaped by human actions (for example, political events) and material forces (like volcanic eruptions), and their interplay (epidemic disease, natural disasters). Latin American texts invite reflection upon mechanisms of enforcement, coercion, violence, conformity, transformation, resistance, and resilience by multiple actors, from those agents of various hegemonic ideologies to the creators and communicators of diversely constructed knowledges of human relationships to the material world. Incorporating gender as a category of importance acknowledges difference and heterogeneity, and can challenge ecocritical scholars and readers to consider more inclusively human experiences and their expressions in knowledge and culture. Informed as it is by both critical gender studies and critical race theory, intersectional ecocriticism undertakes the important task of underscoring a sense of contingency, plurality, and interrelationship in its interpretation of texts.

The pages that follow give an introduction to the field of intersectional ecocriticism as it relates to texts from Latin America, most particularly from areas whose historical realities were shaped by having formed part of the Spanish empire (according to the scope of this volume and of my expertise). Throughout what follows, I adhere to an expansive definition of cultural texts; assert the necessity of foregrounding questions of positionality and intersectionality in Latin American eco-cultural studies; and advocate for critical examination of the material and social dynamics that shape the circulation of texts in a world that is pluriversal but whose institutions have been constructed according to the norms and assumptions of dominant groups. I begin with an exploration of theoretical tools, from work in feminist and postcolonial epistemology to new materialisms, that are available for reconsidering cultural production from diverse moments, places, and creators in Latin America such that questions of environmental imagination and gender politics come to the foreground of study. In the section on "Critical Studies and New Directions", I highlight how existing scholarship, which is informed by theories about gender and place in Latin American contexts, interrogates figures, tropes, and motifs in texts from different periods of literary history. I also point to areas in which there are opportunities for further research. Throughout, I argue for the importance of the recovery and examination of texts that are created in or imagine transitional, liminal moments or spaces, where multiple ideologies come into contact, power shifts, and territories (the body politic, the body human, and bio-geo-physical bodies) are contested or reconstituted in collective imaginations. By way of conclusion, I consider how intersectional ecocritical scholarship in pluricultural contexts like that of Latin America is a critical research practice of importance in illuminating differences and variations in human experiences and communities that popularized discourses about the Anthropocene obscure.

Theoretical Underpinnings for an Intersectional Latin American Ecocriticism

A variety of theoretical and methodological tools facilitate research in intersectional ecocriticism in the Latin American context. Importantly, intersectional Latin American ecocriticism is informed by and contributes to research in various fields in the humanities and social sciences that have attempted to shift academic meaning-making away from "the hubris of the zero point" (Castro-Gómez 2005; Castro-Gómez 2007), an epistemic position anchored in patriarchy, Western modernity, and institutions (universities, disciplines, presses) that have been constructed historically as a consequence of this epistemic position. As shifts in meaning-making happen (and in resistance to shifting and change), new insights and challenges come into focus, as postcolonial and materialist ecocriticism demonstrate, and scholarship begins to reshape disciplines, languages, and frameworks to account for what has been previously underrepresented or excluded from institutional spaces. Foundational articles, chapters, and monographs, including work in feminist ecocriticism, ecofeminism, and masculinity studies published about Latin American texts, suggest the field of inquiry will be a dynamic one.

With what tools do (and might) scholars with an interest in gender, cultural heterogeneity, and ecocriticism study texts and dialogue about them such that the research they produce acknowledges "more inclusive epistemologies" and "more diverse ontologies", as Escobar has urged in political ecology and anthropology (2008, p. 6).[4] The theoretical underpinnings and conceptual tools for intersectional ecocritical studies of Latin American texts are drawn from a heterogeneous array of disciplines, as are its methods. Some of these fields include feminist and postcolonial philosophy; new materialisms and political ecology in anthropology; gender, labor studies, and environmental history; various fields in religious studies, including liberation theology and feminist theology; transdisciplinary and undisciplinary work; and multiple interpretative schools in literary studies. Intersectional ecocritical work examines a variety of cultural products, from published literary texts to films, performances, and digital works. To examine the creation, reception, and circulation of texts, intersectional

---

4   Escobar, Arturo: *Territories of Difference*. Duke University Press: Durham 2008.

ecocriticism turns to archives and other records such as those accessed by digital tools in media studies. In short, as in many other fields of work that grapple with complexity and heterogeneity, each practitioner may draw from an assortment of theories and methods, applying theory to the praxis of interpreting a text or being led from the reading of diverse cultural phenomena to theories (combinations of theories, adaptations of theories, or the creation of new ones) that help tease out their significance.

Of particular relevance to intersectional ecocritical studies is inquiry by historians, anthropologists, and sociologists into gender, power, and place in the Latin American context. Research gathered by historians William E. French and Katherine Elaine Bliss in the edited volume *Gender, Sexuality, and Power in Latin America Since Independence* (2006), for example, draws upon the work of Joan Wallach Scott and asserts that gender is constructed in a dynamic of power, that it "has a history", and that its history is contained in "the social representation of perceived biological differences" that vary in different times and places (p. 1).[5] Because discourse is "the production of particular kinds of knowledge about a subject through the use of language, images, spaces, and symbols" (p. 1), texts can reveal the ways in which categories and identities are mutually constructed. Though the essays are not explicitly environmental in their orientation, the volume includes essays for which analysis is anchored in the importance of place, from enclave economies to urban environments. Importantly, the theoretical orientation that French and Bliss give to the study of gender and power in Latin American history, particularly that gender is "a crucial site or field where power is articulated" (p. 2), has particular relevance for an intersectional ecocriticism involving Latin American texts. Intersectional ecocritical literary and cultural studies scholarship, grounded in the theoretical work like that of French and Bliss and their contributors, will probe the intersections of gender with race, ethnicity, class, and geopolitical location as categories that are mutually contingent and created in relation to one another in all sorts of texts and cultural practices. In its most inclusive praxis, an intersectional ecocriticism can also seek to recover texts that convey epistemologies and imaginaries in which the above categories—of gender, ethnicity, human and non-human agency—simply do not exist in the same ways.

Work published in the first two decades of the twenty-first century in anthropology and sociology by scholars working in Latin American contexts has

---

5   French, William E. / Bliss, Katherine Elaine (eds.): *Gender, Sexuality, and Power in Latin America since Independence*. Rowman & Littlefield Publishers: New York 2006.

introduced new practices to acknowledge and disrupt hierarchies in the creation of scholarship. The works of Arturo Escobar (2008; 2015) and Marisol de la Cadena (2015), which bring the dynamics of power, place, identity, and knowledge, into the foreground, are examples of such research in critical anthropology. Their nuanced theoretical frameworks, marked by an understanding of pluricultural dynamics at multiple levels of scale, is particularly relevant to an intersectional ecocritical practice in a Latin American context. So, too, is work in political ecology, such as the edited volume, *Political Ecology across Spaces, Scales, and Social Groups* (2005) by Susan Paulson and Lisa L. Gezon and in development studies that acknowledge the gendered nature of asset gaps, particularly with regard to land (Barbosa 2015; Deere and León DeLeal 2003; Deere and León DeLeal 2014).

In addition to work anchored in a particular discipline, of relevance to an intersectional ecocriticism, is an increasing number of broadly integrative works of inquiry that might be broadly categorized as works in environmental studies (particularly, environmental humanities and social sciences). This scholarship considers environmental realities in colonial, postcolonial, and neocolonial-extractivist contexts in the global South worldwide. Important contributions include the work of ecological economist Joan Martínez Alier (2003); physicist and activist Vandana Shiva (2005); philosopher Enrique Leff (2002); and theologian Leonardo Boff (1997). Analyses from integrative fields like media studies have yet to include many studies anchored in Latin America or the global South, but as the field becomes more inclusive, it is likely to probe narratives and aesthetics associated with migration, landscape, and the transmission, modification, termination, or transplanting of place-based knowledge; and how knowledges, practices, and exchanges shape the construction of gender categories and roles.

In literary and cultural studies, there are works that are reframing ecocriticism as a whole as a pluralistic field of inquiry and challenging practices for what has often been studied in the field. Stacy Alaimo (2010), for example, makes the case for grappling with trans-corporeality, in which "the human is always intermeshed with the more-than-human world" (p. 2) and argues for an imagination of the context for ethics comprising "the emergent, ultimately unmappable landscapes of interacting biological, climatic, economic, and political forces" (p. 2).[6] Rob Nixon's *Slow Violence and the Environmentalism of the Poor* (2011) calls for rethinking "politically, imaginatively, and theoretically" the

---

6   Alaimo, Stacy: *Bodily Natures: Science, Environment, and the Material Self*. Indiana University Press: Bloomington 2010.

"slow violence" of "delayed destruction that is dispersed across time and space" (p. 2).[7] In feminist ecocriticism, the work of Greta Gaard (1997; 2003; 2011; 2017) has long been a point of reference, as has the work Gaard has coedited with others, such *Ecofeminist Literary Criticism* (1998) with Patrick D. Murphy and *International Perspectives in Feminist Ecocriticism* (2013) with Simon C. Estok and Serpil Oppermann. Catriona Sandilands also has done extensive work involving gender and environment (1999; 2001; 2002). Queer ecocritical studies include works like Catriona Mortimer-Sandilands and Bruce Erickson's edited volume *Queer Ecologies: Sex, Nature, Politics, Desire* (2010) and Nicole Seymour's *Strange Natures* (2013). Additionally, the *Oxford Handbook of Ecocriticism* edited by Greg Garrard (2014) contains some essays on feminist and queer ecocritical perspectives.

In the Latin American context, there are few monographs at present, and a limited number of articles, that make explicit their theoretical orientation in gender studies and ecocritical studies. However, it is likely that the increasing interest in intersectionality in gender studies, the presence of decolonizing practices in multiple disciplines, and the participation in the field of ecocriticism of scholars from diverse backgrounds and/or working a diverse corpus of texts will lead to an abidance of new studies in the years to come. In the following section, therefore, I give a broad overview of existing scholarship that might be either categorized as feminist ecocriticism, ecofeminist criticism, or intersectional ecocriticism or be of importance to work in those areas. Thus, I will include studies that reference gender as a category of analysis but for which the focus is on another aspect of identity, such as race or ethnicity, as well as mention multiple areas for further inquiry for an emergent intersectional practice.

Critical Studies and New Directions

The relative underrepresentation in ecocriticism to date of theoretical approaches grounded in Latin American texts and contexts represents a major opportunity for Latin Americanist scholars to contribute to the field and to bring increasingly into vision the long arc of colonial processes and their operations in relation to gender, ethnicity, place, and sense of place. There is a growing body of work published in Latin American ecocritical studies, in English, by scholars in the United States and Europe, and there have also been important books in ecocriticism published in Spanish in Latin America (Heffes 2013) and Spain

---

7 Nixon, Rob: *Slow Violence and the Environmentalism of the Poor*. Harvard University Press: Cambridge, Massachusetts 2011.

(Flys Junquera, Marrero Henríquez and Barella 2010). Already such works are reframing existing categories and methods of analysis. For example, Gisela Heffes asserts that the realities of Latin America demand a bio/ecocentric approach (2014, p. 32) and an understanding of the dynamics of cultural heterogeneity and intersectionality (pp. 19–21).[8] Chiyo Crawford calls for and models an "anti-colonial feminist environmental justice analysis" (2013, p. 88) in her analysis of the work of a Mexican writer in New York.[9] Work in masculinity studies, such as that of Vinodh Venkatesh (2015), reframes "the masculine" as "a fluid, sociohistorically specific, and interrelational identity" (p. 3).[10] Future scholarship about Latin American texts will build upon existing research in multiple areas of inquiry, and challenge ecocritical scholars to continue to decolonize the field by acknowledging positionality and intersectionality in the generation of theories and methods for analysis as well as in interpretations of primary texts.

In the sections that follow, I first mention monographs and edited volumes that advance feminist ecocriticism, gendered perspectives on ecocriticism, and/or intersectional approaches to ecocriticism in Latin American contexts and then reference examples related to particular periods of literary history.[11] Among monographs and edited volumes, there are several to mention. Ileana Rodríguez is the author of foundational studies like *House/Garden/Nation: Space, Gender, and Ethnicity in Post-Colonial Latin American Literatures by Women* (1994) and *Transatlantic Topographies: Islands, Highlands, Jungles* (2004). Beatriz Rivera-Barnes and Jerry Hoeg in their coauthored book *Reading and Writing the Latin American Landscape* (2009) include a chapter with an explicitly ecofeminist approach, and among the essays in Adrian Kane's edited volume *The Natural World in Latin American Literatures* (2010) are articles grounded in ecofeminism and environmental justice. My ecocritical monograph *Ecological Imaginations in Latin America* (2011) includes gender as a category of analysis

---

8   Heffes, Gisela: "Introducción. Para Una Ecocrítica Latinoamericana: Entre La Postulación de Un Ecocentrismo Crítico y La Crítica a Un Antropocentrismo Hegemónico". *Revista de Crítica Literaria Latinoamericana* 40(79), 2014, pp. 11–34.
9   Crawford, Chiyo; "'Streams of Violence: Colonialism, Modernization, and Gender in María Cristina Mena's 'John of God, the Water-Carrier'". In: Gaard, Greta / Estok, Simon C. / Opperman, Serpil (eds.): *International Perspectives in Feminist Ecocriticism*. Routledge: New York, 2013, pp. 87–100.
10 Venkatesh, Vinodh: *The Body as Capital: Masculinities in Contemporary Latin American Fiction*. University of Arizona Press: Tucson 2015.
11 The overview offered here is far from exhaustive; it is intended to give the reader a foothold for further research in a heterogenous and expanding field of inquiry.

in the consideration of novels published in the last three decades of the twentieth century in Latin America, as does Gisela Heffes's *Políticas de la destrucción / poéticas de la preservación* (2013). Scott DeVries's *A History of Ecology and Environmentalism in Spanish American Literature* (2013) notes the important connection between ecological feminism and environmental justice concerns in the Latin American context. Ecofeminism and gender studies are also fields of reference for multiple contributors to Mark Anderson and Zélia Bora's *Ecological Crisis and Cultural Representation in Latin America* (2016).

In addition to monographs and edited collections, there is an increasing number of journals dedicated to the study of literature and the environment worldwide. *Ecozon@. European Journal of literature, Culture and Environment* is an important outlet for Spanish language ecocriticism and it also admits articles written in English, French, German, and Italian. Journals focused on literary, cultural, and media studies have also published special issues focused exclusively on Latin American ecocriticism or ecocriticism from the global South. Most of these special issues have been guest edited by Latin Americanists in the second decade of the twenty-first century, though a dedicated edition of *Ixquic*, edited by Jorge Paredes and Benjamin McLean, was published in 2000. *Review: Literature and Arts of the Americas* (2012), edited by ecocritic and translator Steven White, was dedicated to ecocriticism in a Latin American context and included an article about pedagogy, ecofeminism, coauthored by Sofia Kearns and me. *Revista de estudios literarios latinoamericanos* (2014), edited by Gisela Heffes, also featured a broad range of essays in Latin American ecocriticism, and various articles, including Heffes's introduction to the volume, underscored the imperative of accounting for heterogeneity in the reading of Latin American texts. The special issue of *Studies in Twentieth and Twenty-First Century Literature* (2015) on eco-testimonial literature from the global South, edited by Erin Finzer, included articles about woman-authored eco-testimonio from Latin America. *Interdisciplinary Studies in Literature and the Environment* featured a special section on Chilean literature (2016). *Millars* (2016) edited by Jerry Hoeg includes articles focused on posthumanism and gender (Martín 2016) and the intersectional nature of networked (digital) activism in the Northern Triangle of Central America (Barbas-Rhoden 2016).

The paragraphs that follow guide readers toward both existing scholarship and questions for further exploration by very broad literary period. I begin with the colonial era, not from lack of recognition of rich cultural production by indigenous groups prior to (and after) 1492 but because few written works from indigenous cultures survived the violence of the colonial era. Primary texts produced during the colonial era often grapple overtly with ontological, epistemological, and phenomenological heterogeneity from the long period of Spanish rule in Latin America (1492 through the early, mid, or late nineteenth century, depending upon territory in question). During this time, diverse indigenous communities and cultural groups in the Americas experienced an advancing frontier of colonial forces of control whose nature was shaped by (and shaped) a long arc of (androcentric, ethnocentric) European intellectual and cultural history—Renaissance, Reformation and Counterreformation, Enlightenment— and the justifications and rhetorics associated with conquest reflect this reality. Anthropologist Arturo Escobar underscores that there is a "coloniality of nature in modernity that needs to be unveiled" (2008, p. 8);[12] with this unveiling must also come a consideration of how the mechanisms of coloniality and modernity have been gendered acts that perform, reify, and enforce social roles and acts with material consequences. The expanding colonial frontier continually produced contested spaces in which multiple ideologies, knowledge systems, and ways of being came into contact and conflict in the body politic, human bodies, and bio-geo-physical bodies over a long span of time in a vast territory, and intersectional ecocritical inquiry into colonial era texts is thus potentially a vast and heterogeneous area of study.

Studies like Jennifer French's "Naturaleza y subjetividades en la América Latina colonial" (2014) and Ileana Rodríguez's *Transatlantic Topographies* (2004) suggest promising areas of research for ecocritical inquiry into coloniality. A return to the archives can also yield insights into colonial imaginations of gender, ethnicity, and the material world. For example, Federico Garza Carvajal examines records related to the prosecution of "sodomites" in *Butterflies Will Burn* (2010) and demonstrates how perceptions of manliness intertwine with discourses and mechanisms of control in early modern Spain and New Spain. His work examines the ideological writings of the *moralistas* and, in particular, their teachings on manly conduct and the transmission of ideological frameworks in which men are understood as agents of continual creation, and collaborators with God, because their seed harbors the potential for future beings (p. 17).[13] Further inquiry into primary texts, both in circulation and in archives, will likely continue to explore gender roles in relationship to diverse world views, physical places and territories, and emergent hierarchies of class and ethnicity; the regulation and transgression of identities and subjectivities related to gender, desire,

---

12 Escobar, Arturo: *Territories of Difference*. Duke University Press: Durham 2008.
13 Garza Carvajal, Federico: *Butterflies Will Burn: Prosecuting Sodomites in Early Modern Spain and Mexico*. University of Texas Press: Austin 2010.

and place; as well as genealogies of resistance, coercion, cooperation, subversion, subterfuge, and violence on the part of heterogeneous groups in the colonial milieu.

Primary texts from the era of the imagination of independence, struggles for independence, and period of nation formation are likewise promising for intersectional ecocritical approaches. The power differentials of colonialism continued beyond independence, as the governments of nascent Latin American republics pursued neocolonial practices for the subjugation of indigenous groups and other marginalized communities living within the political borders of the new republics. In this area, future ecocritical scholarly inquiry will build upon important groundwork in the study of gender and nation; nation and place; and place and ethnicity in both Latin American and comparative American contexts. Doris Sommer's classic *Foundational Fictions* (1993) is an early work of relevance, as is Mary Louise Pratt's *Imperial Eyes* (1991).

There is some existing scholarship on gender and place that suggests a promising future for explicitly ecocritical practice with regard to texts associated with nation formation. For example, *Race Mixture in Nineteenth-Century U.S. and Spanish American Fictions: Gender, Culture, and Nation Building* (2004) by Debra J. Rosenthal considers how lives and desires unfolded at the intersections of imperial expansion, national political organization, and new avenues of trade and wealth accumulation throughout the Americas. Vinodh Venkatesh in *The Body as Capital* (2015) focuses on male virility, corporal aesthetics, and science in *caudillo* novels since the nineteenth century (p. 6). Future intersectional ecocritical work might examine the intersectional nature of the construction (and exclusion) of knowledge in institutional (educational, governmental) and civic settings (lecture societies or research associations, for example) and the influence of positivism, eugenics, and theories of race associated with nineteenth century European imperialism in the rhetoric of foundational texts in the early republics. Recurring tropes from the era of independence and nation formation can also be examined with an intersectional ecocritical lens, including the romantic repurposing of heroes of indigenous resistance from the colonial past and the ways they are privileged or denigrated in gendered terms. Intersectional ecocriticism may also make fruitful inquiry into gendered tropes associated with urbanization and institutionalization (the lettered city, the lettered gentleman, the naturalist, the amateur archaeologist), hierarchies of privilege and disdain in association with particular landscapes (coasts, highlands, lowlands, forests) and the people who inhabit them, and the foundational ways in which the emergence of metropolitan culture underwritten by commodity booms shaped cultural production. Literary modernism itself is a propitious field for intersectional ecocritical readings of its cultivated pleasurescapes.

Inquiry on twentieth century literary texts and cultural expressions is the area in which there are the most numerous studies in Latin American ecocriticism *per se*; gender is a category of critical analysis in many, though it is a primary focus in fewer. For an ecocritical approach to texts from the early twentieth century, such as *novelas de la tierra* and regionalist writing, Jennifer French's volume *Nature, Neo-Colonialism and the Spanish-American Regional Writers* (2005) is a foundational monograph. Erin Finzer has authored ecofeminist analyses of the work of early twentieth century women writers like Gabriela Mistral, Magda Portal, and Romelia Alarcón de Folgar (2009; 2015a; 2015b; 2015c; 2015d; 2015e). Though its focus is not specifically on either gender or ecocriticism, Ericka Beckman's *Capital Fictions* (2013) signals the importance of considering the intersection of the realities that are economic, material, and gendered, especially as she inquires into the changing patterns of consumption as export economies were established in Latin America. Future intersectional ecocritical analyses of texts from the mid-twentieth century, such as those associated with the Boom, may yield new readings of oft-research writers like Gabriel García Márquez, Isabel Allende, and Carlos Fuentes, as well as authors less frequently studied. The cultural milieu shaped by revolution is also favorable for intersectional ecocritical exploration of musical culture (Victor Jara, Violeta Parra, the Mejía Godoy brothers), writings in multiple genres by activists and militants, testimonies of genocide and civil war, and documentary film.

With regard to the late twentieth century and early twenty-first century, from the late 1980s forward, there are three important trends in intersectional ecocriticism (as well as other types of criticism): challenges within the academy to canons, an increased accessibility of texts with the advent of the Internet, and a fragmented editorial landscape throughout the Spanish-speaking world. These conditions facilitate the pursuit of intersectional ecocritical practice and make the endeavor more complex. In order to be responsive to Latin American cultural realities, an intersectional ecocriticism of late twentieth and twenty-first century primary texts must be inclusive of texts and cultural practices sited in urban areas, as well as rural ones. Ecocritical practice must include in its scope of inquiry transnational texts and imaginaries. Texts in this vein include transnational and diasporic ones featuring dislocations, deterritorializations, discontinuities, and fragmentation. It should also interrogate the ways migration, displacement, and deterritorialization disrupt the transmission of embedded knowledge across generations, as well as generate new means to understand being-in-place. Intersectional ecocriticism must also consider new "testimonial" genres, like digital narratives, citizen journalism, and manifestos by grassroots activists and collective entities, many of them led by women and composed of politically marginalized groups on the front lines of the resistance to extractivist operations. It must examine texts authored by indigenous, Afro-Latin American, and immigrant and diasporic writers, and consider cultures of habitat and mechanisms of deterritorialization over a long arc of history.

There is substantial research into late twentieth and early twenty-first century texts that takes a feminist ecocritical, ecofeminist, and/or intersectional approach to reading primary texts. Sofía Kearns's (1998) essay on Central American author Anacristina Rossi is among the earliest such pieces, and she is also the author of a study on the ecopoetics of numerous women poets (2006). I have several articles informed by feminist ecocriticism, including studies on works of Gioconda Belli and Anacristina Rossi (Barbas-Rhoden 2005); Libertad Demitrópulos and Sylvia Iparraguirre (Barbas-Rhoden 2008); and Tatiana Lobo (Barbas-Rhoden 2010); and another with a focus on queer identities, mobility, and the regulation of space in two texts set in Nicaragua (Barbas-Rhoden 2013). Regina Root has an important translation project involving Anacristina Rossi's ecotestimonial novel, *La loca de Gandoca*, the first chapter of which appears in *International Perspectives in Feminist Ecocriticism* (2013). *Rain Forest Literatures* (2004) by Lucia Sá includes gender and other aspects of identity in its analysis of Amazonian texts. *Caribbean Literature and the Environment* (2005), edited by Elizabeth M. DeLoughrey, Renée K. Gossen, and George B. Handley, features contributions informed by postcolonial perspectives on gender and race in the Caribbean. Additionally, DeLoughrey and Handley's introduction to the volume *Postcolonial Ecologies* (2011) is an important contribution that explicitly seeks to recover ecofeminist and postcolonial work in the genealogy of ecocriticism. Priscilla Ybarra's *Writing the Goodlife: Mexican American Literature and the Environment* (2016) is a seminal work on one rich area of Latinx literary production in the United States, and her work has long had a focus on gender and environmental justice. Salma Monani and Joni Adamson likewise have opened a significant space for dialogue in *Ecocriticism and Indigenous Studies* (2016).

Film and media studies will continue to develop as an important area of inquiry. Amanda Holmes's studies (2006; 2012; 2017), for example, are important points of reference for ecocritical film studies work attentive to place, identity, and politics. Work on documentary and feature film, like Sharada Balachandran Orihuela and Andrew Carl Hageman's study of gender and (US/ Mexican) border ecologies in documentaries (2011), and Jorge Marcone's consideration of "more-than-environmental social movements" (2015, p. 209) as depicted in film, draws attention to the intersectional nature of environmental injustices.[14] There are also important forthcoming ecocritical volumes that will expand the scope of scholarship: one edited by Carolyn Fornoff and Gisela Heffes on film, and another by Ana Maria Mutis, Elizabeth Pettinaroli, and Ilka Kressner on *Ecofiction and Ecorealities* which will be inclusive of work in media studies, broadly defined.

Conclusion

By way of conclusion, it is worth underscoring that intersectional ecocriticism about Latin American texts is likely to continue to develop as integrative and interdisciplinary scholarship. Rather than fitting neatly into categories ideated in other parts of the world, in other languages, it will exist in dialogue with scholarship informed by texts written in other places and from other geopolitical positions. Expanding the range of cultural texts studied in ecocriticism and foregrounding questions of intersectionality in the analysis of those texts, especially those from a pluricultural region like Latin America, illuminates differences and variations in human experiences and communities that are of crucial importance in an era in which people worldwide grapple with what it means to live, survive, and hope in a changing material world. Practicing intersectional ecocriticism will also nudge the environmental humanities, as they are institutionalized in diverse departments, institutes, conferences, and journals worldwide becoming more pluralistic and marked by scholarship that signals its own premises, privileges, and assumptions. In the case of the study of Latin American cultural texts, this change means a shifting from "zero points" that have dominated discourse about place: patriarchal viewpoints, colonialist, ethnocentric, monolinguistic, and heteronormative ones. The understanding of ecocriticism will change and gender studies will change. From the moments of learning and unlearning we ecocritics have with one another, with the students we teach, and the public with which we engage there may emerge more nuanced understandings of who people are and of our diverse ways of imagining and being in the world.

---

14 Marcone, Jorge: "Filming the Emergence of Popular Environmentalism in Latin America". *Global Ecologies and the Environmental Humanities: Postcolonial Approaches* 41, 2015, p. 207.

Bibliography

Alaimo, Stacy: *Bodily Natures: Science, Environment, and the Material Self.* Indiana University Press: Bloomington 2010.

Anderson, Mark, and Zelia M. Bora (eds.): *Ecological Crisis and Cultural Representation in Latin America: Ecocritical Perspectives on Art, Film, and Literature.* Lexington Books: Lanham 2016.

Barbas-Rhoden, Laura: "Activismo medioambiental multimodal en el Triángulo Norte de Centroamérica: medios digitales, patrimonio biocultural y de-colonialidad". *Millars* 40(1), 2016, pp. 155–178. doi:http://dx.doi.org/10.6035/Millars.2016.40.9.

Barbas-Rhoden, Laura: "Biopolitics and the Critique of Neoliberalism in El Corazón Del Silencio by Tatiana Lobo". *A Contracorriente* 8(1), 2010, pp. 259–276.

Barbas-Rhoden, Laura: *Ecological Imaginations in Latin American Fiction.* University Press of Florida: Gainesville 2011.

Barbas-Rhoden, Laura: "Ecology, Coloniality, Modernity: Argentine Fictions of Tierra Del Fuego". *Mosaic: An Interdisciplinary Critical Journal* 41(1), 2008, pp. 1–18.

Barbas-Rhoden, Laura: "Espacio, violencia y heteronormatividad en una Nicaragua transnacional: Lectura ecofeminista de *La Yuma* y *Meet Me Under the Ceiba.*" In: Ortiz Wallner, Alexandra / Mónica Albizúrez Gil (eds.): *Poéticas y políticas del género.* Walter Frei: Berlin 2013.

Barbas-Rhoden, Laura: "Greening Central American Literature". *Interdisciplinary Studies in Literature and Environment*, 2005, pp. 1–17.

Barbosa, Luiz C.: *Guardians of the Brazilian Amazon Rainforest: Environmental Organizations and Development.* Routledge: New York 2015.

Beckman, Ericka: *Capital Fictions: The Literature of Latin America's Export Age.* University of Minnesota Press: Minneapolis 2013.

Boff, Leonardo: *Cry of the Earth, Cry of the Poor.* New York: Orbis Books, 1997.

Castro-Gómez, Santiago: "La hybris del punto cero." In Castro-Gómez, Santiago / Grosfoguel, Ramón (eds.): *El giro decolonial.* Pontificia Universidad Javeriana: Bogotá 2005, pp. 79–92.

Castro-Gómez, Santiago: "The Missing Chapter of Empire: Postmodern Reorganization of Coloniality and Post-Fordist Capitalism". *Cultural Studies* 21(2–3), 2007, pp. 428–448.

Crawford, Chiyo: "'Streams of Violence: Colonialism, Modernization, and Gender in María Cristina Mena's 'John of God, the Water-Carrier'". In: Gaard,

Greta / Estok, Simon C. / Opperman, Serpil (eds.): *International Perspectives in Feminist Ecocriticism*. Routledge: New York, 2013, pp. 87–100.

Crenshaw, Kimberle: "Demarginalizing the Intersection of Race and Sex: A Black Feminist Critique of Antidiscrimination Doctrine, Feminist Theory and Antiracist Politics". *The University of Chicago Legal Forum* 140, 1989, pp. 139–167.

Deere, Carmen Diana / León de Leal, Magdalena (eds.): *Empowering Women: Land and Property Rights in Latin America*. University of Pittsburgh Press: Pittsburgh 2014.

Deere, Carmen Diana (eds.): "The Gender Asset Gap: Land in Latin America". *World Development* 31(6), 2003, pp. 925–947.

De la Cadena, Marisol: *Earth Beings: Ecologies of Practice across Andean Worlds*. Duke University Press: Durham 2015.

DeLoughrey, Elizabeth / Didur, Jill / Carrigan, Anthony (eds.): *Global Ecologies and the Environmental Humanities: Postcolonial Approaches*. Routledge: New York 2015.

DeLoughrey, Elizabeth M. / Gossen, Renee K. / Handley, George B. (eds.): *Caribbean Literature and the Environment: Between Nature and Culture*. University of Virginia Press: Charlottesville 2005.

DeLoughrey, Elizabeth / Handley, George B. (eds.): *Postcolonial Ecologies: Literatures of the Environment*. Oxford University Press: New York 2011.

DeVries, Scott M.: *A History of Ecology and Environmentalism in Spanish American Literature*. Rowman & Littlefield: New York 2013.

Escobar, Arturo: *Territories of Difference*. Duke University Press: Durham 2008.

Escobar, Arturo: "Territorios de Diferencia: La Ontología Política de Los 'Derechos Al Territorio'". *Cuadernos de Antropología Social* 41, 2015, pp. 25–38.

Finzer, E. S.: "Bleeding Mud: The Testimonial Poetry of Hurricane Mitch in Nicaragua". *Studies in 20th & 21st Century Literature* 39(2), Jan. 2015a. doi:10.4148/2334-4415.1838.

Finzer, E. S.: "La Endometriosis, El Exceso y El Periodo Especial En La Poesía de Reina María Rodriguez". *Letras Femeninas* 35(2), 2009.

Finzer, E. S.: "Grafting the Maya World Tree: Cosmic Conservation in Romelia Alarcón de Folgar's Llamaradas (Guatemala, 1938)". *Isle: Interdisciplinary Studies in Literature and Environment*, 22(2), 2015b.

Finzer, E. S.: "Mother Earth, Earth Mother: Gabriela Mistral as an Early Ecofeminist". *Hispania: A Journal Devoted to the Teaching of Spanish and Portuguese* 98(2), 2015c.

Finzer, E. S.: "Putting Environmental Injustice on the Map: Ecotestimonies from the Global South". *Studies in 20th & 21st Century Literature* 39(2), 2015d, doi:10.4148/2334-4415.1837.

Finzer, E. S.: "Trees, Seas, and Ecofeminist Imaginary in the Vanguard Poetry of Magda Portal". *Hispanófila* 173, 2015e.

Flys Junquera, Carmen / Marrero Henríquez, José Manuel / Barella Vigal, Julia (eds.): *Ecocríticas, Literatura y medio ambiente*. Iberoamericana Vervuert, 2010.

French, William E., and Katherine Elaine Bliss (eds.): *Gender, Sexuality, and Power in Latin America since Independence*. Rowman & Littlefield Publishers: New York 2006.

French, Jennifer L: *Nature, Neo-Colonialism and the Spanish-American Regional Writers*. Dartmouth University Press: Hanover, 2005.

French, Jennifer L: "Naturaleza y Subjetividades En La América Latina Colonial: Identidades, Epistemologías, Corporalidades". *Revista de Crítica Literaria Latinoamericana* 40(79), 2014, pp. 35–56.

Gaard, Greta: *Critical Ecofeminism*. Lexington Books: Lanham 2017.

Gaard, Greta: *Ecofeminism and Globalization: Exploring Culture, Context, and Religion*. Rowman & Littlefield Publishers: Lanham 2003.

Gaard, Greta: "Ecofeminism Revisited: Rejecting Essentialism and Re-Placing Species in a Material Feminist Environmentalism". *Feminist Formations* vol. 23(2), 2011, pp. 26–53.

Gaard, Greta: "Toward a Queer Ecofeminism". *Hypatia* 12(1), 1997, pp. 114–137.

Gaard, Greta Claire / Murphy, Patrick D. (eds.): *Ecofeminist Literary Criticism: Theory, Interpretation, Pedagogy*. University of Illinois Press: Illinois 1998.

Gaard, Greta / Estok, Simon C. / Opperman, Serpil (eds.): *International Perspectives in Feminist Ecocriticism*. Routledge: New York 2013.

Garrard, Greg: *The Oxford Handbook of Ecocriticism*. Oxford Handbooks: Oxford 2014.

Garza Carvajal, Federico: *Butterflies Will Burn: Prosecuting Sodomites in Early Modern Spain and Mexico*. University of Texas Press: Austin 2010.

Heffes, Gisela: *Políticas de La Destrucción, Poéticas de La Preservación: Apuntes Para Una Lectura (Eco) Crítica Del Medio Ambiente En América Latina*. Beatriz Viterbo: Buenos Aires 2013.

Heffes, Gisela: "Introducción. Para Una Ecocrítica Latinoamericana: Entre La Postulación de Un Ecocentrismo Crítico y La Crítica a Un Antropocentrismo Hegemónico". *Revista de Crítica Literaria Latinoamericana* 40(79), 2014, pp. 11–34.

Holmes, Amanda. "Introduction: Identity Maps of Hispanic Cinema". *Revista Canadiense de Estudios Hispánicos*, 2012, pp. 1–8.

Holmes, Amanda. *Politics of Architecture in Contemporary Argentine Cinema*. Palgrave Macmillan: New York 2017.

Holmes, Amanda. "Regarding Gender and Nature in" Madeinusa"(2006)". *Revista Canadiense de Estudios Hispánicos*, 2012, pp. 203–216.

Kane, Adrian Taylor: *The Natural World in Latin American Literatures: Ecocritical Essays on Twentieth Century Writings*. McFarland: Jefferson, North Carolina 2010.

Kearns, Sofía: "Nueva Conciencia Ecológica En Algunos Textos Femeninos Contemporáneos". *Latin American Literary Review* 34(67), 2006, pp. 111–127.

Kearns, Sofía: "Otra Cara de Costa Rica a Través de Un Testimonio Ecofeminista". *Hispanic Journal*, 1998, pp. 313–339.

Leff, Enrique: *Saber ambiental: sustentabilidad, racionalidad, complejidad, poder*. Siglo XXI: Mexico 2002.

Lugones, María: "Toward a Decolonial Feminism." *Hypatia* 25(4), 2010, pp. 742–759.

Marcone, Jorge: "Filming the Emergence of Popular Environmentalism in Latin America". *Global Ecologies and the Environmental Humanities: Postcolonial Approaches* 41, 2015, p. 207.

Martín, Juan Carlos: "El planeta hembra de Gabriela Bustelo: Descifrando una identidad poshumana". *Millars* 40(1) 2016, pp. 81–97. doi:http://dx.doi.org/10.6035/Millars.2016.40.5.

Martínez-Alier, Joan: *The Environmentalism of the Poor: A Study of Ecological Conflicts and Valuation*. Edward Elgar Publishing: Northampton 2003.

Mignolo, Walter: *The Darker Side of Western Modernity: Global Futures, Decolonial Options*. Duke University Press: Durham 2011a.

Mignolo, Walter: "Epistemic Disobedience, Independent Thought and Decolonial Freedom". *Theory, Culture & Society* 26(7–8), 2009, pp. 159–181. doi:10.1177/0263276409349275.

Mignolo, Walter: "Geopolitics of Sensing and Knowing: On (de) Coloniality, Border Thinking and Epistemic Disobedience". *Postcolonial Studies* 14(3), 2011b, pp. 273–283.

Monani, Salma / Adamson, Joni (eds.): *Ecocriticism and Indigenous Studies: Conversations from Earth to Cosmos*. Routledge: New York 2016.

Mortimer-Sandilands, Catriona / Erickson, Bruce (eds.): *Queer Ecologies: Sex, Nature, Politics, Desire*. Indiana University Press: Bloomington 2010.

Nixon, Rob: *Slow Violence and the Environmentalism of the Poor.* Harvard: Cambridge, Massachusetts 2011.

Oppermann, Serpil: "Feminist Ecocriticism: A Posthumanist Direction in Ecocritical Trajectory." In: Gaard, Greta / Estok, Simon C. / Opperman, Serpil (eds): *International Perspectives in Feminist Ecocriticism.* Routledge, 2013, pp. 19–36.

Orihuela, Sharada Balachandran / Hageman, Andrew Carl: "The Virtual Realities of US/Mexico Border Ecologies in Maquilapolis and Sleep Dealer". *Environmental Communication* 5(2), 2011, pp. 166–86. doi:10.1080/17524032.2011.565063.

Paulson, Susan / Gezon, Lisa L., (eds.): *Political Ecology across Spaces, Scales, and Social Groups.* Rutgers University Press, 2005.

Pereyra, Marisa: "Paradise Lost: A Reading of Waslala from the Perspectives of Feminist Utopianism and Ecofeminism". *The Natural World in Latin American Literatures: Ecocritical Essays on Twentieth Century Writings*, 2010, pp. 136–153.

Pratt, Mary Louise: *Imperial Eyes: Travel Writing and Transculturation.* Routledge: New York 1992.

Rivera-Barnes, Beatriz / Hoeg, Jerry (eds.): *Reading and Writing the Latin American Landscape.* Palgrave Macmillan: New York 2009.

Rodríguez, Ileana: *Transatlantic Topographies: Islands, Highlands, Jungles.* University of Minnesota Press: Minneapolis 2004.

Rodríguez, Ileana: *Women, Guerrillas, and Love.* University of Minnesota Press: Minneapolis 1996.

Root, Regina: "Saving the Costa Rican Rainforest: Anacristina Rossi's *Mad About Gandoca*". In: Gaard, Greta / Estok, Simon C. / Opperman, Serpil (eds.): *International Perspectives in Feminist Ecocriticism.* Routledge, 2013. pp. 101–119.

Rosenthal, Debra J.: *Race Mixture in Nineteenth-Century U.S. and Spanish American Fictions : Gender, Culture, and Nation Building.* The University of North Carolina Press: Raleigh 2004.

Sá, Lúcia: *Rain Forest Literatures: Amazonian Texts and Latin American Culture.* University of Minnesota Press: Minneapolis 2004.

Sandilands, Catriona: "Desiring Nature, Queering Ethics". *Environmental Ethics* 23(2), 2001, pp. 169–188.

Sandilands, Catriona: *The Good-Natured Feminist: Ecofeminism and the Quest for Democracy.* U of Minnesota Press: Minneapolis 1999.

Sandilands, Catriona: "Lesbian Separatist Communities and the Experience of Nature: Toward a Queer Ecology". *Organization & Environment* 15(2), 2002, pp. 131–163.

Seymour, Nicole: *Strange Natures: Futurity, Empathy, and the Queer Ecological Imagination*. University of Illinois Press: Champaign, Illinois 2013.

Shiva, Vandana: *Earth Democracy: Justice, Sustainability and Peace*. Zed Books: London 2005.

Sommer, Doris: *Foundational Fictions: The National Romances of Latin America*. University of California Press: Berkeley 1991.

Venkatesh, Vinodh: *The Body as Capital: Masculinities in Contemporary Latin American Fiction*. University of Arizona Press: Tucson 2015.

Ybarra, Priscilla Solis: *Writing the Goodlife: Mexican American Literature and the Environment*. University of Arizona Press: Tucson 2016.

Spanish Ecocriticism

Pamela Phillips
Enlightening Nature: An Ecocritical Reading of Eighteenth-Century Spanish Literature*

Abstract: A common thread in the traditional history of Spanish literature marginalizes the eighteenth century and values its aesthetic sensibility towards place insofar as it announces the romantic gaze. As for the first tendency, there is no doubt that recent scholarship has secured the texts of the 1700s their deserved place in the literary chronology. The romanticization of eighteenth-century aesthetics has delayed the recognition of the value of specifically that period's writing on the natural world and society's relationship to it on its own terms, and not through the lens of a posterior mindset. Nature is an integral part of Enlightenment thought; indeed, the view of nature as something to be conquered or controlled by man dates to this period, as well as the posterior recognition of nature as a limited resource. Much eighteenth-century Spanish writing deals with nature and environmental concerns in ways that are less egocentric than the Romantic model and much more in line with contemporary models, thus making it a valuable resource for the ecocritical dialogue in general and for the local case of Spain in the twenty-first century. This chapter responds to ecocriticism's intermittent attention to eighteenth-century Spanish literature. The ecological reconsideration of eighteenth-century Spanish literature will examine canonical texts such as Feijoo's "Honra y provecho de la agricultura", Jovellanos's "Descripción del Castillo de Bellver" and his "Informe sobre la ley agraria", to name a few, as demonstrations of a profound concern for the natural world, its ecological, economic, and moral meaning, and the relationship between the human and the non-human world. Likewise, reading a selection of the period's poetry proves to be a useful source to uncover the ecological consciousness and concerns of eighteenth-century Spain.

Keywords: Ecocriticism, Spanish Literature, Eighteenth-Century Spain

At first glance, the proposal to practice ecocriticism on eighteenth-century Spanish literature may lift more than one eyebrow from different subject disciplines. Ecocriticism's strong ties to British Romantic studies established an unofficial timeline that left eighteenth-century letters on the sidelines. The traditional narrative on the European Enlightenment assigns Spain the role of passive recipient of the intellectual innovations generated by its northern neighbors. Finally, an inquiry into its literature from a similar perspective classifies the period's writing as an inferior parenthesis between its Golden Age predecessors

---

* My thanks go to Dayanira Moya Pérez for providing me with valuable research assistance.

and its Romantic followers. Fortunately, the present-day exercises of rethinking the Enlightenment in a global context, on the one hand, and ecocriticism, on the other, are successfully challenging the established interpretations of these areas of knowledge. As a result, the Spanish Enlightenment is recognized as one of the many national Enlightenments, and eighteenth-century literary studies and ecocriticism are slowly growing closer.[1] This chapter capitalizes on these points of confluence to bring eighteenth-century Spanish writing into the mainstream discussion of ecocriticism, enriching by extension both fields of knowledge.

The members of the eighteenth-century Spanish Republic of Letters had much to say about their natural surroundings. Although they were by no means unique in their viewpoint, Benito Jerónimo Feijoo, Gaspar Melchor de Jovellanos, Antonio Ponz, Pedro Rodríguez Campomanes, and Juan Meléndez Valdés, among other principal observers of eighteenth-century Spain, display an awareness and sensitivity to the value of nature and to those human behaviors that affect continuously and cumulatively their kingdom's terrain. Practicing ecocriticism on their writing brings these texts out of their traditional discipline in search of new levels of meaning until now not considered. Likewise, this literary archive can contribute considerably to ecocriticism. Nature and the environment are contemporary concerns that acquire greater relevance when historicized. What and why we think as we do today about nature and the natural world have a lot to do with the eighteenth century. Among the many texts awaiting ecocritical inquiry are works from eighteenth-century Spain, as their authors experienced, described, and evaluated nature and a diversified range of very current issues that include public health, land and water management, biodiversity, pollution, demography, deforestation, among others. Inscribed in the scenes that captured their attention and with which they filled their poems, travel accounts, discourses, and official reports are questions related to the social and economic policies of eighteenth-century Spain, and that contemporary ecological and political circles and beyond continue to tackle. To read

---

1 Jesús Astigarraga's edited volume *The Spanish Enlightenment revisited* defends a place for Spain in the European Enlightenment on its own terms. Mariselle Meléndez and Karen Stolley's "Introduction: Enlightenments in Ibero-America" offers a careful review of Enlightenment scholarship's coverage of the Spanish-American Enlightenment that contributes to the ongoing revision of the global eighteenth century. As I will document further on, Erin Drew, David Fairer, and Christopher Hitt stand out among the scholars of eighteenth-century English literature who have successfully inserted their field into ecocritical studies.

eighteenth-century Spanish literature through the prism of ecocriticism is thus to acknowledge one more "local" scenery in the fruitful critical exchange taking place at the greater "global" level.

A brief sketch, even if summarily, of the history of ecocriticism yields insight into the significance and complexity of engaging with eighteenth-century Spanish literature through this critical approach. The multiple definitions of ecocriticism coincide in it being the study of the representation of the natural world in literary texts on its own terms and as a point of departure to consider its relationship with humankind.[2] One of the fastest growing fields of academic study, ecocritical scholarship marks its evolution in waves whose currents intermingle in a palimpsest fashion (Buell 2005, p. 17).[3] The result is a diverse field that has expanded in multiple academic directions, languages, time periods, and geographies.[4] Ecocriticism traces its foundation to studies in late eighteenth-century English poetry and the already-present Romantic mindset, and from there it fastens on to that period's pastoral tradition and Anglo-American nature writing. In an exponential manner, ecocritics' fascination with and concern for the natural world have led to groundbreaking readings of not only Romantic texts, but also early modern literature and contemporary works in different languages. As for the specific case of literature in Spanish, the visibility and treatment of the natural world in Hispanic-American and Caribbean writing have made it much more receptive to ecocritical inquiry than Spanish letters (Marrero Henríquez 2010).[5] Despite this lopsidedness, ecocriticism has established itself as a recognized field of study in Hispanic Studies academic circles, as evidenced by the upsurge in scholarship and active research units. A series of recent publications is slowly bridging the gap between the Spanish and Spanish-American literary

---

2   Although this most inclusive definition of ecocriticism embraces animal studies, this chapter will confine itself exclusively to the treatment of the natural world in eighteenth-century Spain. In progress is a future publication that will secure a place for eighteenth-century Spanish studies in the scholarly framework for thinking about representations of animals in literature.
3   Buell, Lawrence: *The Future of Environmental Criticism: Environmental Crisis and Literary Imagination*. Blackwell Publishing: Oxford 2005.
4   For detailed reviews of the history of ecocriticism, see Buell, Lawrence / Heise, Ursula K. / Thornber, Karen: "Literature and Environment". *Annual Review of Environment and Resources* 36, 2011, pp. 417–440. See also Marland, Pippa: "Ecocriticism". *Literature Compass* 10/11, 2013, pp. 846–868.
5   Marrero Henríquez, José Manuel. "Ecocrítica e Hispanismo". In: Flys Junquera, Carmen / Marrero Henríquez, José Manuel / Barella Vigal, Julia (eds.): *Ecocríticas. Literatura y medio ambiente*. Iberoamericana/Vervuert: Madrid 2010, pp. 193–217.

production, but the role of the eighteenth-century contribution has yet to gain a foothold in this effort to green Spanish letters.[6]

The roots of ecocriticism's underexamination of eighteenth-century Spanish letters can be traced to the intersection of the history of the period's thinking about nature and the emergence of the ecological crisis. As Lorraine Daston affirms, nature lies at the center of the familiar narration of the Enlightenment project: "For the enlightened, nature was the principle that unified all narratives— the history of human society, as much as the history of the earth and stars, was a narrative about nature" (p. 503).[7] During this period, Western thinking about the natural world underwent a significant transformation. Alongside the inherited providential interpretation of nature surged an awareness of nature's limitations and the role human agency plays in its constitution and longevity. Spain's intellectual community was receptive to this dialogue. Luis Urteaga refers to eighteenth-century Spanish thinking about the natural world as a philosophical and cultural kaleidoscope in which classical, Aristotelian, and Neoplatonic ideas mix with scholasticism and the new scientific practices (p. 15).[8] No single event in the eighteenth century better confirmed the entanglement of these lines of thinking as the 1755 Lisbon earthquake.[9] Inspired by the ideals of Enlightenment, the gradual process of modernization that ensued entailed human intervention in the physical environment in multiple ways, ranging from the clearing of trees for grazing and other purposes, swamp draining, and myriad land use practices. These actions were carried out in the spirit of progress, an end that justified the mastery of nature. This focus on improvement through the use of nature fueled the period's anthropocentric turn and what would become a thorn in ecocritical

---

6   See José Manuel Marrero Henríquez and Julia Barella Vigal's contributions to the anthology *Ecocríticas*. Dolores Thion Soriano-Mollá's collection overlooks the treatment of landscape in eighteenth-century Spanish letters, and Dale Pratt and Barbara Gordon's entry dedicated to Spain in Patrick Murphy's *Literature of Nature: An International Sourcebook* focuses exclusively on the nineteenth- and twentieth-century literature.
7   Daston, Lorraine. "Afterword: The Ethos of Enlightenment". In: William Clark/Jan Golinski/Simon Schama (eds.): *The Sciences in Enlightened Europe*. University of Chicago Press: Chicago & London, 1999, pp. 495–504.
8   Urteaga, Luis: "Explotación y conservación de la naturaleza en el pensamiento ilustrado". *Geocrítica* 50, 1984, pp. 7–46.
9   See Braun, Theodore / Radner, John B. (eds.): *The Lisbon Earthquake of 1755: Representations and Reactions*. Voltaire Foundation: Oxford, 2005. See also Ordaz, Jorge: "Monográfica: El terremoto lisboeta de 1755". *Cuadernos Dieciochistas* 6, 2005, pp. 19–247.

studies. Nowhere is the resulting Enlightenment-Romanticism tension better summed up than by Jonathan Bates in his *Foreword* to Laurence Coupe's *The Green Studies Reader*: "The starting point of Laurence Coupe's *The Green Studies Reader* is the Romantic critique of the Enlightenment's aspiration to master the natural world and set all things to work for the benefit of human commerce" (Fairer, p. 203).[10] This kind of critical appreciation has contributed to the marginalization of eighteenth-century writing in the ecocritical dialogue, a scholarly exchange that, in David Fairer's opinion, has become "myopic" (p. 203).

In response and as a corrective to this critical nearsightedness, recent scholarship on eighteenth-century English literature by David Fairer, Christopher Hitt, and Erin Drew has succeeded in improving ecocritical regard for this written material, thus making it of special interest in the search for an ecocritical conceptual framework to study Spanish letters of the period. Hitt makes the point that eighteenth-century representations of nature are ambivalent: "For the great legacy of that age with respect to attitudes about nature is best described [...] as] the paradoxical recognition that we both master and are mastered by the nonhuman world" (2004, p. 126).[11] For Erin Drew, usufruct offers a methodological frame of reference to scrutinize eighteenth-century English literature on its own terms.[12] Likewise, David Fairer's reading of Georgic poetry invigorates English studies and tweaks ecocriticism's chronology to secure a place for eighteenth-century writing. In making a case for an ecocritical assessment of eighteenth-century literature, this scholarly corpus foregrounds the importance of contextualization.[13] There is no denying the eighteenth century's anthropocentrism, but a return to the definitions of ecocriticism reminds us that wrapped up in this critical category is human agency. It is virtually impossible to study and consider nature in isolation of human intervention. Keeping in mind the emphasis on the interconnections between the human and the natural world serves to modify the critique of the eighteenth-century attitude towards the

---

10 Fairer, David: "'Where Fuming Trees Refresh the Thirsty Air': The World of Eco-Georgic". *Studies in Eighteenth-Century Culture* 40, 2011, pp. 201–218.
11 Hitt, Christopher: "Ecocriticism and the Long Eighteenth Century". *College Literature* 31 (3), 2004, pp. 123–147.
12 Drew, Erin: "'Tis Prudence To Prevent Th'Entire Decay": Usufruct and Environmental Thought". *Eighteenth-Century Studies* 49 (2), 2016, pp. 195–210.
13 It is perhaps no coincidence that Jesús Astigarraga argues in favor of the contextual approach as a means to draw eighteenth-century Spain into the mainstream Enlightenment discussion (pp. 10–11). In: Astigarraga, Jesús (ed.): *The Spanish Enlightenment revisited*. Voltaire Foundation: Oxford 2015, pp. 1–17.

latter. The focus on *use* would have made it seem less attractive for the ecocritical gaze fixed on Romantic and nature writing, but it is intriguing to survey how the eighteenth century thought about its relationship with land and rural space. Taking care and utility as its cue, this chapter examines the way prominent voices of the period turn Spain into a text, reading in its fields, forests, and landscape the problems, errors, and intents to progress humankind and nurture its natural surroundings.

Eighteenth-century Spain was primarily a rural space whose economy was bound to agriculture.[14] Plots of land were a site of labor that needed to be controlled, organized, and managed as part of the program to strengthen the kingdom's economy and well-being. In essence, much of the period's writing asks and responds to the question "what to do with the land?" Although the central administration sponsored various initiatives to elaborate agrarian reform proposals, these efforts did not reach the implementation stage. The political and social climate in eighteenth-century Spain did not welcome the implications of a new agrarian design, as historical scholarship has documented; nevertheless, this literature is a valuable source to understand the thinking on the topic.[15] At work in these texts is the projection of an urban vision of the rural space through the eyes of observers who were not farmers nor were they trained economists. Many, like Campomanes, Jovellanos, Ponz, and others, came into contact with nature through their travels, an exercise that trained their eyes to capture the reality of the kingdom. Writing about their travels offered them a distinct mode of expression to engage directly with that natural space and publicize it. Jovellanos's domestic tours prepared him to draw up his *Report on [...] Agrarian Law* (*Informe de la Sociedad Económica de Madrid al Real y Supremo Consejo de Castilla en el Expediente de Ley Agraria*).[16] Others, like Benito Jerónimo Feijoo and Juan Meléndez Valdés, used the essay and poetry, respectively, to respond to ecological changes and their implications for their country's well-being. While they cannot be considered environmentalists in the contemporary sense of the word, insofar as their perception of the natural world is grounded in the Enlightenment's emphasis on reason and utility, their writing brings out their

---

14 See Martí, Marc: *Ciudad y campo en la España de la Ilustración*. Milenio: Lleida, 2001, pp. 27–124.
15 See the essays included in García Sanz, Ángel / Sanz Fernández, Jesús (eds.): *Reformas y políticas agrarias en la historia de España (De la Ilustración al primer franquismo)*. Ministerio de Agricultura, Pesca y Alimentación; Madrid 1996, pp. 15–200.
16 All translations from Spanish to English within this chapter are mine.

admiration and concern for this space. The act of reading then helps move their expressions from sentiment to the possibility of action.

Included in volume 8 of his *Teatro crítico universal*, published in April 1739, Feijoo's "Honra y provecho de la agricultura" ["Honor and Benefit of Agriculture"] was instrumental in focusing attention on agriculture and its agents. Taking the geography of Asturias, Galicia, and the northern portion of León as his point of reference, Feijoo raises his voice against the declining agrarian production: "El descuido de España lloro, porque el descuido de España me duele" (p. 451) "[I cry for Spain's neglect, because Spain's neglect hurts me"]. As the title of the essay advances, Feijoo invests agriculture and the farmer with honor and identifies practices and habits that warrant correction in the name of improvement.[17] Farming is a noble service, Feijoo argues, because it owes itself to a superior force: "de todas las demás artes fueron autores los hombres; de la agricultura lo fue Dios" (p. 441) ["men were the author of all the other arts; God was the creator of agriculture"]. In his opinion, Spain suffers from an inversion of priorities that has favored military strength over agrarian prosperity. Feijoo uses the metaphor of illness, diagnosing the kingdom with gout, to spotlight the decline in agrarian activity and productivity. The remedy, according to Feijoo in the role of physician, is to reorient the kingdom's human resources towards agriculture.

Motivating the writing is an effort to correct the false notion that Spain's natural resources were abundant. Already in the first half of the century it was clear that the real panorama was very different. Resources were limited and the lives of those most linked to the land were far from bucolic, as Feijoo impresses upon: "¿Pero hay hoy gente más infeliz, que los pobres labradores? ¿Qué especie de calamidad hay, que aquéllos no padezcan?" (p. 460) ["But are there people today unhappier than the poor laborers? What type of calamity do they not suffer?"]. "Honra y provecho de la agricultura" epitomizes the eighteenth-century exaltation of agriculture as a motor of prosperity, but with its call to attention to the dire existence of the rural population and obvious social inequalities, the text acquires ecocritical relevance, resonating with contemporary thinking about farming and the rural culture.[18]

We can hear echoes of Feijoo's "Honra y provecho..." some fifty years later in Antonio Ponz's endorsement of agriculture as the base of the kingdom's wealth

---

17 See Fallows, Noel: "Farms and Letters: Feijoo's «Honra y provecho de la agricultura»". *Revista de Estudios Hispánicos* 30 (2), 1996, pp. 285–295.
18 William Major argues in favor of expanding the reach of ecocriticism to include agrarianism. See Major, William: "The Agrarian Vision and Ecocriticism". *Interdisciplinary Studies in Literature and Environment* 14 (2), 2007, pp. 51–70.

and stability: "la primera de las artes es la nobilísima agricultura, sin cuyo estado floreciente, o desaparecerán las nobles artes, las ciencias y toda industria, o caerán en languidez a proporción de su decadencia" (XII, vi, 14) ["the leader of the arts is the most noble agriculture, without whose flourishing state the noble arts, the sciences and all industry will disappear, or they will fall listless in proportion to their decadence"].[19] This profound respect for agriculture surfaces again in the transcription of a letter from the traveler's acquaintance "el señor N***" [Mr. N***] in the last pages of the multi-volume *Viaje*: "cuanto más floreciente fuese la agricultura, tanto más poderoso y fuerte será el Estado en que se ejerza" (XVIII, vi, 34) ["the more plentiful may be agriculture, the more powerful and strong will be the State in which it is carried out"].

The eighteenth-century Spanish Republic of Letters shared its European contemporaries' enthusiasm for the aesthetics of nature. This should come as no surprise as the *tertulias*, or private discussion groups, were a principal space that stimulated new aesthetic perspectives on nature. The gathering sponsored by Pablo de Olavide in Seville during the years 1767 and 1776, for example, was the point of entry of Alexander Pope's poetry, a major influence on the way Spanish writers would look and write about rural landscapes (Urteaga, p. 24).[20] Likewise, the Georgic stress on work and productivity offers a model for the design of eighteenth-century Spanish landscape descriptions.[21] Throughout the second half of the century, a number of poems celebrate the virtues of the natural world. "Égloga en alabanza de la vida en el campo" ["Eclogue in Praise of Country Life"] won Juan Meléndez Valdés first place in the Royal Academy of the Spanish Language's 1780 competition for the best eclogue that celebrated rural life and Tomás de Iriarte was awarded the second prize for his poem "La felicidad de la vida en el campo" ["Happiness of Life in the Country"].[22] Another poem by Meléndez Valdés, "Romance XV: Los segadores" ["The Harvesters"], is of interest

---

19 Ponz, Antonio. *Viaje de España*. 4 Volumes. Aguilar: Madrid, 1988.
20 Urteaga, Luis. "Explotación y conservación de la naturaleza en el pensamiento ilustrado". *Geocrítica* 50, 1984, pp. 7–46.
21 Casalduero and García Calderón discuss the Georgic influence in their respective essays. Casalduero, Joaquín: "Las nuevas ideas económicas sobre la agricultura en el siglo dieciocho y el nuevo sentimiento de la naturaleza". *Estudios de literatura española*. Editorial Gredos: Madrid, 1973, pp. 172-185. García Calderón, Ángeles: "La poesía inglesa de la naturaleza en el XVIII y su influencia en Meléndez Valdés". *Revista de Literatura* LXIX (138), 2007, pp. 519–541.
22 For coverage of the competition and analysis of the eclogues, see Martí, pp. 193–229.

for framing its portrayal of rural Spain in the motif of useful beauty.[23] The bulk of the poem is an ode pronounced by Plácido, an aging farmer striving to motivate his crew of harvesters as they go out to work in the fields at dawn. Laced throughout Plácido's call to action are key ingredients of the period's reform program. The poem foregrounds the topics of hard work, discipline, and expertise through a sketch of Edenic toil, not a struggle with nature. The harvesters' youth gives them the strength Plácido no longer has (vv. 49–60). Work is celebrated as a deterrent to laziness and wrongdoing, thus reminding the reader of Charles III's 1783 declaration that work is good and honorable: "el trabajo es pasatiempo/cuando el placer lo acompaña" (vv. 75–6) ["work is a hobby/when pleasure accompanies it"]. Reading the poem ecocritically identifies farm work as the place and activity where humankind and nature connect. Just as Feijoo expressed his concern for the plight of the rural poor, Plácido instills in his crew empathy for those not as fortunate to work (vv. 101–105). In a didactic turn, the poem ends with an image of the farmers following Plácido to the fields, as if they were students following their teacher (vv. 153–156). The bucolic fields described in the opening stanzas are not a backdrop, but rather are interconnected with human needs and real problems.[24] At work in Plácido's ode is a celebration of the fields as a natural resource and a national prize that secures the kingdom's stability.

The awe, reverence, and pride that Spain's nature aroused in the principal voices of the period are also strikingly visible in other genres. A seaside stroll infuses Jovellanos with a solemn respect for nature and its power over humankind, as he records in his *Diario*:

> No puedo echar de mi memoria la situación de *Santa Catalina* en la noche de ayer. La dudosa y triste luz del cielo; la extensión del mar, descubierta de tiempo en tiempo por medrosos relámpagos que rompían el lejano horizonte; el ruido sordo de las aguas, quebrantadas entre las peñas al pie de la montaña; la soledad, la calma y el silencio de todos los vivientes hacían la situación solemne y magnífica […] interrumpió mis meditaciones el ¿*Quién vive*? De un centinela apostado en un pórtico de una ermita, […] y esta única voz, de que yo me alejaba poco a poco, contrastaba maravillosamente con el silencio universal. ¡Hombre!, si quieres ser venturoso, contempla la Naturaleza, y

---

23 Meléndez Valdés, Juan: "[Romance XV] Los segadores". In: Reyes, Rogelio (ed.): *Poesía española del siglo XVIII*. Cátedra: Madrid 2011, pp. 255–259.
24 For a reading of the treatment of rural space in Meléndez Valdés's poetry, see Calvo Revilla, Joaquín: "El nuevo sentido del campo en la poesía de Meléndez". *Ínsula* 179, 1961, p. 6.

acércate a ella; en ella está la fuente del escaso placer y felicidad que fueron dados a tu ser. (2008–2010, vol. 6, pp. 621–622)

[I cannot get out of my memory the situation of *Santa Catalina* last night. The doubtful and sad light of the sky; the extension of the sea, discovered from time to time by the timid lightning that broke the distant horizon; the rumbling sound of the water, shattering between the rocks at the foot of the mountain; the solitude, the calm and the silence of all the living made the situation solemn and magnificent (...) a *Who goes there?* of a guard stationed in the chapel entrance interrupted my meditations. And this sole voice, from which I moved away little by little, contrasted marvelously with the universal silence. Indeed! If you want to be fortunate, contemplate Nature, get close to her; in her is the source of the little pleasure and happiness given to you].[25]

Quoting this passage extensively highlights the priority given to the physical sensations and sensory engagement with the surroundings. Throughout his *Diario*, Jovellanos transmits his delight upon beholding scenes of sublimity through exclamations, the repetitive use of the superlative, and winks to the *locus amoenus*:[26] A rest under a hazel tree in San Andrés de Trubia during his trip from Gijón to León in June 1792 arouses a state of bliss: "¡Oh naturaleza! ¡Qué desdichados son los que no pueden disfrutarte en estas augustísimas escenas, donde despliegas tan magníficamente tus bellezas y ostentas toda tu majestad!" (2008–2010, vol. 6, pp. 392–395) ["Oh nature! How unfortunate are those who cannot enjoy you in these impressive scenes, where you unfold magnificently your beauty and flaunt your majesty!"].[27] The expression of gratitude towards the divine power contained in celebrations like these of nature's wonders is suggestive of the meshing of the providential vision with anthropocentrism: the natural world serves the human one in a harmonious way that ultimately owes itself to the superior divine forces.[28] In making a case for an ecological assessment of the sublime aesthetic, Christopher Hitt argues that "Part of the sublime experience,

---

25 Jovellanos, Gaspar Melchor de: *Obras completas*. 14 vols. Editorial KRK: Oviedo, 2008–2010.
26 Jovellanos records numerous moments of the sublime experience in his private writing, as Ana Rueda and Elena de Lorenzo Álvarez have studied. See Rueda, Ana: "Jovellanos en sus escritos íntimos: el paisaje y la emoción estética de 'lo sublime'". *Revista de Literatura* LXVIII (136), 2006, pp. 489–502; and Lorenzo Álvarez, Elena de: "El curioso contemplador de la naturaleza: la estética de lo sublime en los escritos literarios de G. M. de Jovellanos". *Iberoromania* 84, 2016, pp. 270–280.
27 Similar exclamations are documented throughout his *Diario*; see, for example, Jovellanos 2008–2010, vol. 7, pp. 247–249, and vol. 9, pp. 483–486.
28 See Sitter, John: "Eighteenth-Century Ecological Poetry and Ecotheology". *Religion & Literature* 40 (1), 2008, pp. 11–37.

in other words, is the realization that we are mortal creatures, "beings of nature" whose lives are entirely dependent on forces greater than we are" (1999, p. 607). Reading these passages from Jovellanos's *Diario* ecocritically thus brings to light the blurred line between natural and human history in the eighteenth century.

Scenes of nature's abundance and fertility soften the difficulties of traveling through eighteenth-century Spain and single out those places as conducive to the kingdom's progress. A good example of this equation is Ponz's reaction to the spectacle in the Mancha: "Terrenos áridos y pelados sequerales no producen ideas de gusto y alegría" (XVI, i, 84) ["Dry fields and bare plots do not produce tasteful and joyful ideas"]. Throughout the *Viaje de España*, Ponz uses *pelado* or barrenness as synonymous with overall deterioration, while an enjoyable or appealing landscape is *divertido*. Natural beauty played a role in the creation of a positive image of the kingdom in the eyes of its neighbors. Extensive tracts of barren land like those along the route to Toledo, for example, contribute to the negative perception many foreigners had of Spain (I, i, 2). The chestnut trees, orchards, and vineyards surrounding the Duke's Palace in the village of Béjar were pleasing to the traveler's eye, contrasting sharply to the nearby bare banks of the Guadiana and Evora rivers (VIII, i, 15; VIII, v, 23). Drawing on theories of affect and emotion, Ponz goes one step further, comparing the state of nature to a thermometer of society's inclination to improvement. The unremitting traveler associates nature's fertility with civility and social well-being; in contrast, arid and barren extensions condition its inhabitants' mindset to refrain from agricultural initiative. In this way, scenes of beauty and their unpleasant counterparts offer themselves to eighteenth-century Spanish observers as sites for their improvement programs. Insofar as agriculture serves as a measure of the kingdom's well-being, writers-improvers represent the natural world in terms of its potential for progress or development. The green world thus asserts a greater significance in this literary production as it becomes a major force in the direction of eighteenth-century Spanish political thought, exercising a sort of activism, adjusted to the political and intellectual atmosphere of the time.

Intertwined in eighteenth-century Spanish writers' celebration of nature's beauty is a discourse of care and concern for that very space. Their gaze values a bucolic landscape, but it also focuses in on the hardships of the terrain. If agriculture is a driving force in the kingdom's improvement, beauty is utility in eighteenth-century Spanish land evaluation. At issue here is the relationship between humankind and nature, and for eighteenth-century Spanish observers, the kingdom's progress and prosperity depended upon its rational and responsible use of the land. The eighteenth-century mindset perceives the relationship between humankind and nature as dynamic and interconnected: insofar as humans depend on nature, it is their responsibility to respect and care for it. A useful point of entry to understand the period's writing, care is a critical category that presupposes an understanding of the natural world, as William Major writes: it is "therefore an intrinsically radical concept as it asks that we understand the world in a certain way, guided by personal contact with the earth and the values learned therein" (p. 64).[29] The discourse of care, although complex because it cannot be easily detached from questions of power and dependency, offers an understanding of the eighteenth-century relationship with nature that attenuates the anthropocentric label. Again, William Major's argument is pertinent: "The challenge is to employ best practices to achieve more harmonious relations between the two rather than ignore one of the fundamental truths about what it is to be human: that we must work, that we must use, and that all we do necessarily impacts nature, for good or ill" (pp. 62–63). Richard Kerridge also makes a case for "care" that aligns eighteenth-century Spanish literature with ecocriticism's focus: "'Care' preserves the range of possibilities, from incremental and gradually spreading change to abrupt social revolution. The word encompasses feeling ("caring about") and action ("take care of")" (p. 365).[30] The trick is to gauge the breaking point. It becomes a question of proper and controlled exploitation in order to sustain natural resources for the nation's well-being.

This new perspective of the natural world derives from and nurtures a dynamic intellectual environment that contributed to impulse the sciences in Spain during the eighteenth century. The production and circulation of knowledge in this period is located in specific spaces directly linked to the state and the public domain. Reaching back to the foundation of the Royal Academy of Mathematics in 1582, the effort to promote the sciences in Spain enjoyed its most intense progress during the second half of the century, specifically the reigns of Ferdinand VI (1746–1759) and Charles III (1759–1788). The sciences found support in the creation of Madrid's Royal Botanical Garden and the Real Cabinet of Natural History, the Royal Institute and Observatory of the Spanish Navy in Cádiz, the different local chapters of the Economic Societies of Friends of the Country, and the Royal Asturian Institute, among other official institutions. The list is much longer, but what it reveals is an active participation in the international scientific

---

29 Major, William: "The Agrarian Vision and Ecocriticism". *Interdisciplinary Studies in Literature and Environment* 14 (2), Summer 2007, pp. 51–70.
30 Kerridge, Richard: "Ecocritical Approaches to Literary Form and Genre: Urgency, Depth, Provisionality, Temporality". In: Garrard, Greg (ed.): *The Oxford Handbook of Ecocriticism*. Oxford University Press: New York, 2014, pp. 361–76.

dialogue and one that extended beyond the royal court and Madrid to other points of the kingdom.[31] Access to scientific literature was another stimulus to this new line of thinking. Madrid-based printing houses distributed the Spanish translation of Pluche's *Spectacle de la Nature* (9 volumes, 1732–1742) between 1753 and 1785, the same year in which José Clavijo y Fajardo's translation of Buffon's *Historia Natural* began to circulate. With his *Introducción a la Historia Natural, y a la Geografía Física de España* Guillermo Bowles contributes to the Bourbon kingdom's study of its natural history, but he calls on Spaniards to take a closer look at their country (p. 1). Another essential ingredient that affected the appreciation of the natural world were the numerous scientific expeditions, whose coverage is beyond the scope of this chapter.[32] Finally, the history of "naturalista" also sheds light on the materialization of ecological thinking in Spain. The word appears in the 1734 edition of the Royal Academy of the Spanish Language's *Dictionary* with a scientific meaning: "el que trata, averigua y examina las virtudes, propiedades y calidades de los entes naturales, especialmente de los animales, plantas, minerales" (p. 652) ["a person who deals with, discerns and examines the virtues, properties, and qualities of natural beings, especially animals, plants, minerals"]. What emerges from this map of the institutional situation is proof that Spain, like its neighbors, engaged with Enlightenment scientific knowledge.

The institutional advances were significant, but they were not enough. Over the course of the century, there was also a constant call to insert agriculture in the Enlightenment pedagogical project. During the first half of the century, Feijoo urged the need for more technical literature (pp. 455–457). In keeping with his mission to correct common errors, Feijoo takes issue with the habit of plowing with mules rather than oxen, a much more economical and efficient animal (pp. 472–476). The Benedictine monk also celebrates the advantages of yoking oxen at the neck, for both the farmer and the beast. Years later, during his tour of Andalusia, Ponz was adamant about the need to open up the kingdom's universities to the formal study of agriculture as a corrector to inefficient habits passed down through the generations, like those that Feijoo had drawn attention to earlier in the century (XVI, iii, 26; XVIII, i, 83–84). It was not enough, Antonio Ponz argued in the Prologue to Volume XIII of his *Viaje de España*, to promote tree planting; rather, it was essential to teach how to do it. To that end

---

31 See, for instance, García Camarero, pp. 3–6; Martí, pp. 127–146; Pimentel 2015.
32 See Pimentel, Juan: *Testigos del mundo. Ciencia, literatura y viajes en la Ilustración.* Marcial Pons Historia: Madrid 2003.

the tireless traveler distributed that Prologue as an offprint to the farmers of the kingdom's capital (Blasco Castiñeyra, p. 304).[33] His concern for the well-being of both the farmer and nature comes through in his coverage of the olive harvest in Tortosa, where shaking the trees is not the preferred method: "Hay en el término de Tortosa grandes olivares; pero no sacuden las plantas, como en gran parte de Castilla, para coger el fruto; lo que es un gran disparate, pues con esta bárbara operación no medran los árboles ni dan a sus dueños el fruto tal vez doblado que daría" (XIII, vi, 20) ["There are great olive trees within the Tortosa area; but they do not shake the trees, like in most of Castile, to collect the fruit; this is a great mistake, since with this senseless maneuver the trees do not grow nor do they give their owners half of what they could"].

As these examples show, the discourse of the mastery over nature in eighteenth-century Spanish writing does not disregard the need to protect that same environment. There is a clear awareness that while human life most certainly had a bearing on nature, the opposite was also true: humankind was responsible for nature's longevity. An early voice to speak out in defense of the kingdom's natural resources and also critical of human conduct is that of Fray Martín Sarmiento in 1757. His appraisal deserves to be quoted in full:

> Así yo no necesito recurrir a Revoluciones celestes para palpar las causas de las decadencias de muchos Mixtos en España. Falta carbon y leña; porque se corta, y no se planta. Faltan carnes; porque por ser más regaladas las crías, se comen y se apuran. Falta el pasto; porque faltando ya la leña, se arrancan para la lumbre, hasta las mismas raíces de todo combustible. Faltan los Pescados en el Mar; porque se desprecian las leyes de la veda que se pusieron justamente a favor de la cría. Faltan en los ríos porque con la *cal coca*, *Torvisco* y con otros iniquios medios de pescar se pesca todo de un golpe, y de un golpe se queda el río sin pesca. (Urteaga, p. 28)

> [I do not need to resort to celestial revolutions to detect the causes of the decadence of many *mixtos* in Spain. There is no coal nor firewood, because the trees are felled and not planted. There is a shortage of meat, because since the young animals are a steal, they are eaten and used up. There is a lack of pasture, because given the lack of timber, the grass is pulled out to use for the fire, even the very roots of all fuel. The fish in the sea are missing because there is disrespect for the laws of the ban that was imposed precisely in favor of the young. There are none in the rivers because with the *cal coca*, *Torvisco* and other vile fishing practices, everything is caught at once, and all at once the river is left with no fish].

---

33 Blasco Castiñeyra, Selina. "El «Viaje de España» de don Antonio Ponz. Compendio de las alteraciones introducidas por el autor en todas las ediciones de su obra". *Anales de Historia del Arte* 2, 1990, pp. 223–304.

Likewise, in his 1794 inaugural address to the Royal Asturian Institute, Jovellanos exhorted that human intervention in nature follow a strict code of ethics:

> Pero guardaos, amados compatriotas, de abusar de este precioso instrumento; guardaos de aplicarle a objetos que no sean dignos de su excelencia y nuestra vocación. No olvidemos jamás que nos fue dado para mejorar nuestra existencia y concurrir al bien del género humano, y que si somos llamados al estudio de la naturaleza, no es para satisfacer nuestro orgullo, sino para socorrer nuestra miseria. (1993, p. 399)

> [But take care, fellow citizens, that you do not abuse this precious instrument; beware of applying it to objects that are not worthy of its excellence and our vocation. Don't ever forget that it was given to us to improve our existence and turn everything to humankind's good, and if we are called to study nature, it is not to satisfy our pride, but to come to the aid of our misery].

With these words of caution Jovellanos envisions a human-nature relationship that challenges the anthropocentric discourse of mastery. Humankind depends on nature to satisfy its basic needs, but any overindulgence puts its good standing at risk. Caring for nature, as Jovellanos contends, becomes a mirror of human behavior.

Of the many ecological topics covered in eighteenth-century Spanish writing, forests and enclosure stand out for the passions they aroused. The state of the kingdom's forests generated much writing in the period, and its review, as well as that of the critical scholarship on the topic, is beyond the scope of this chapter.[34] It is important to remember that in addition to Antonio Ponz's spirited arboreal defense in the pages of his *Viaje de España* over the course of some twenty years, Jovellanos speaks out in his 1795 *Report on Agrarian Law*, and Casimiro Gómez Ortega, the botanist and founder of Madrid's Royal Botanic Garden, asked for a translation of Henri-Louis Duhamel du Monceau's treatise *Tratado de las siembras y plantíos de árboles y de su cultivo* (1773). Informing these calls for reforestation is the sentiment of care for the natural world as a prime beneficiary, as Ponz succinctly affirms in the Prologue to Volume IX: "Sin duda que la escasez de árboles causa la sequedad del clima, la esterilidad de la tierra, la falta de granos y otros males" (IX, Prologue, note) ["Without a doubt the scarcity of trees causes the dry climate, the sterility of the land, the lack of grain, and other

---

34 This chapter joins with my reading of eighteenth-century Spanish forest literature to offer an important literary contribution to ecocriticism; see Phillips, Pamela: "Journeys through the Forest in Eighteenth-Century Spain and Spanish America". In: Marrero Henríquez, José Manuel (ed.): *Transatlantic Landscapes: Environmental Awareness, Literature, and the Arts*. Biblioteca Benjamin Franklin: Alcalá de Henares 2016, pp. 29–47.

evils"]. Another important idea that works itself out in this forest writing is the connection between arboreal plenitude and social improvement. The domestic travels of many Spaniards in the eighteenth century positioned them as firsthand observers of the depopulation and the consequential economic decline afflicting the interior provinces of the kingdom. In this regard, Ponz was of the opinion that Spain had sufficient natural resources and capabilities to provide for its citizens and promote demographic growth. Planting trees, he maintained, would spark the recuperation of depressed regions, like Castile.

The power and influence of the Mesta was also signaled out as a primary obstacle to rural Spain's progress. Among the most polemical topics was enclosure as it constituted a direct challenge to the power and existence of the Mesta, the medieval association of transhumant sheep and cattle farmers. Antonio Ponz signaled the Mesta's role in the depopulation of Extremadura: "los montes de encinas y las dehesas [...] son la causa de la infelicidad de esta tierra [...] porque pudiera dársele al territorio cultivo más útil y conducente a la población" (VII, vii, 4, nota 1) ["the oak forests and the meadows ... are the cause this region's unhappiness ... because they could offer the territory a more useful crop"]. Pedro Rodríguez Campomanes also witnessed transhumance's negative effects. In his account of his 1778 trip from Madrid to Extremadura, he defends enclosure as part of a greater agrarian reform: "Estas tierras [Jaraicejo], si no se cercan [...]. no pueden tener progreso, a causa de ser la garganta por donde pasan tres millones de merinos dos veces al año" (p. 198) ["This area (Jaraicejo), if it is not enclosed... cannot progress, because it is the pass through which three million merino sheep cross twice a year"].[35] Campomanes and Ponz speak plainly enough about the process of rural depopulation that resonates very clearly with Spanish current concerns.

The relationship between the natural surroundings and human misconduct surfaces time and again in the discourse of care. The eighteenth-century Spanish gaze admired the population's commitment to well-tended fields. Campomanes, for instance, praises the villagers who live along the border with Portugal: "en lugar de costas merecían premio los que tratan de reducir a cultura unos yermos que lastiman a todo buen ciudadano, que los ve desiertos de hombres y poblados de fieras" (p. 221) ["instead of coastline, prizes are in order for those who try to convert wasteland that distresses all good citizens, who see it abandoned of men and full of fierce-looking animals"]. But along with his contemporaries

---

35 Rodríguez de Campomanes, Pedro: *Viajes por España y Portugal*. Editorial Miraguano: Madrid 2006.

he agrees that in certain circumstances logging and clearing are necessary. Cultivated nature, for example, deters crime: "Para el viaje de la Reina se alineó el camino, que es terreno de arena gruesa y firme, cortando los árboles y ramas que impedían su dirección arreglada. También se *desbrozó el monte de ambos lados, que por su espesura y despoblación era abrigo de ladrones*"(p. 216) ["For the Queen's journey the road was aligned, being a terrain of course and firm gravel, cutting the trees and branches that prevented its orderly direction. Also, the woodland on both sides was cleared because its thickness and depopulation gave shelter to thieves"]. Antonio Ponz observes a similar situation in the Mancha: "Desde la Venta de Cárdenas hasta La Carolina hay quatro leguas [...] cultivan veinte y ocho, o treinta fanegas de tierra cada uno, y algunos más de cincuenta, donde antes nada se cogía, pues todo eran espesuras y matorrales, *abrigos de lobos y de ladrones*" (XVI, ii, 8) ["From the Venta de Cárdenas to Carolina there are four leagues... they farm twenty-eight, or thirty bushels each one, and some more than fifty, where before nobody worked, because it was all thickets and bushes, a refuge for wolves and thieves"]. If a common thread can be discerned in these comments, it is the shared conviction that a degraded physical environment equates to a hazardous human one, and the opposite also holds true. Eighteenth-century Spanish writers take pleasure in beholding a cultivated and tended to natural environment because for them it is a sign of a human one aligned with the enlightened ideals of good taste, progress, and well-being.

By opening up eighteenth-century Spanish literature to an innovative instrument like ecocriticism, it becomes clear that the general anthropocentric conclusions regarding the period are in need of crucial reconsideration. The primary concern for Spain's enlightened circle was the kingdom itself, and its improvement and happiness were directly grounded in nature, specifically, cultivated and controlled land. Through their literary labor not only did they pronounce upon the state of Spain's agricultural scene and forests, but they also recorded their sentimental and aesthetic attachment to these spaces. This selection of eighteenth-century Spanish writing acquires a greater level of meaning when we take into account that it becomes an archive of landscapes that would be inevitably transformed, even erased, as a result of the following century's industrialization and environmental alteration.

Bibliography

Astigarraga, Jesús: "Introduction: *admirer, rougir, imiter*—Spain and the European Enlightenment". In: Astigarraga, Jesús (ed.): *The Spanish Enlightenment revisited*. Voltaire Foundation: Oxford 2015, pp. 1–17.

Barella Vigal, Julia: "Naturaleza y paisaje en la literatura española". In: Carmen Flys Junquera / Marrero Henríquez, José Manuel / Barella Vigal, Julia (eds.): *Ecocríticas: Literatura y medio ambiente*. Iberoamericana/Vervuert: Madrid 2010, pp. 219-238.

Blasco Castiñeyra, Selina: "El *Viaje de España* de don Antonio Ponz. Compendio de las alteraciones introducidas por el autor en todas las ediciones de su obra". *Anales de Historia del Arte* 2, 1990, pp. 223-304.

Bowles, Guillermo: *Introducción a la Historia Natural, y a la Geografía Física de España*. Second Edition, Corrected. Imprenta Real: Madrid 1782.

Braun, Theodore / Radner, John B. (eds.): *The Lisbon Earthquake of 1755: Representations and Reactions*. Voltaire Foundation: Oxford 2005.

Buell, Lawrence: *The Future of Environmental Criticism: Environmental Crisis and Literary Imagination*. Blackwell Publishing: Oxford, 2005.

Buell, Lawrence / Heise, Ursula K. / Thornber, Karen: "Literature and Environment". *Annual Review of Environment and Resources* 36, 2011, pp. 417-40.

Calvo Revilla, Joaquín: "El nuevo sentido del campo en la poesía de Meléndez". *Ínsula* 179, 1961, p. 6.

Casalduero, Joaquín: "Las nuevas ideas económicas sobre la agricultura en el siglo dieciocho y el nuevo sentimiento de la naturaleza". In: *Estudios de literatura española*. Editorial Gredos: Madrid, 1973, pp. 172-185.

Daston, Lorraine: "Afterword: The Ethos of Enlightenment". In: Clark, William / Golinski, Jan / Schama, Simon (eds.): *The Sciences in Enlightened Europe*. University of Chicago Press: Chicago & London 1999, pp. 495-504.

Drew, Erin: ""Tis Prudence To Prevent Th'Entire Decay": Usufruct and Environmental Thought". *Eighteenth-Century Studies* 49 (2), 2016, pp. 195-210.

Fairer, David: "'Where Fuming Trees Refresh the Thirsty Air': The World of Eco-Georgic". *Studies in Eighteenth-Century Culture* 40, 2011, pp. 201-218.

Fallows, Noel: "Farms and Letters: Feijoo's «Honra y provecho de la agricultura»". *Revista de Estudios Hispánicos* 30 (2), 1996, pp. 285-295.

Feijoo, Benito Jerónimo: *Teatro crítico universal*. Editorial Castalia: Madrid 1986.

García Camarero, Ernesto: "La regeneración científica en la España del cambio de siglo". *Revista de Hispanismo Filosófico* 5, 2000, pp. 17-42.

García Calderón, Ángeles: "La poesía inglesa de la naturaleza en el XVIII y su influencia en Meléndez Valdés". *Revista de Literatura* LXIX (138), 2007 July-December, pp. 519-41.

García Sanz, Ángel / Sanz Fernández, Jesús (eds.): *Reformas y políticas agrarias en la historia de España*. Ministerio de Agricultura, Pesca y Alimentación: Madrid 1996.

Glacken, Clarence J.: *Traces on the Rhodian Shore. Nature and Culture in Western Thought from Ancient Times to the End of the Eighteenth Century*. University of California Press: Berkeley, CA 1967.

Hitt, Christopher: "Ecocriticism and the Long Eighteenth Century". *College Literature* 31 (3), Summer 2004, pp. 123–147.

Hitt, Christopher: "Towards an Ecological Sublime". *New Literary History* 30 (3), 1999, pp. 603–623.

Kerridge, Richard: "Ecocritical Approaches to Literary Form and Genre: Urgency, Depth, Provisionality, Temporality". In: Garrard, Greg (ed.): *The Oxford Handbook of Ecocriticism*. Oxford University Press: New York 2014, pp. 361–376.

Jovellanos, Gaspar Melchor de: "Oración inaugural a la apertura del Real Instituto Asturiano". In: Jovellanos, Gaspar Melchor de: *Poesía. Teatro. Prosa Literaria*. Taurus: Madrid 1993, pp. 388–411.

Jovellanos, Gaspar Melchor de: *Obras completas*. 14 vols. Editorial KRK: Oviedo 2008–2010.

Lorenzo Álvarez, Elena de: "El curioso contemplador de la naturaleza: la estética de lo sublime en los escritos literarios de G. M. de Jovellanos". *Iberoromania* 84, 2016, pp. 270–280.

Major, William: "The Agrarian Vision and Ecocriticism". *Interdisciplinary Studies in Literature and Environment* 14 (2), Summer 2007, pp. 51–70.

Marland, Pippa: "Ecocriticism". *Literature Compass* 10/11, 2013, pp. 846–868.

Martí, Marc: *Ciudad y campo en la España de la Ilustración*. Milenio: Lleida 2001.

Marrero Henríquez, José Manuel: "Ecocrítica e Hispanismo". In: Flys Junquera, Carmen / Marrero Henríquez, José Manuel / Barella Vigal, Julia (eds.): *Ecocríticas: Literatura y medio ambiente*. Iberoamericana/Vervuert: Madrid 2010, pp. 193–217.

Meléndez, Mariselle / Stolley, Karen. "Introduction: Enlightenments in Ibero-America". *Colonial Latin American Review* 24 (1), 2015, pp. 1–16.

Meléndez Valdés, Juan: "[Romance XV] Los segadores". In: Reyes, Rogelio (ed.): *Poesía española del siglo XVIII*. Cátedra: Madrid 2011, pp. 255–59.

Ordaz, Jorge (coord.): "Monográfica: El terremoto lisboeta de 1755". *Cuadernos Dieciochistas* 6, 2005, pp. 19–247.

Phillips, Pamela: "Journeys through the Forest in Eighteenth-Century Spain and Spanish America". In: Marrero Henríquez, José Manuel (ed.): *Transatlantic Landscapes: Environmental Awareness, Literature, and the Arts*. Biblioteca Benjamin Franklin: Alcalá de Henares 2016, pp. 29–47.

Pimentel, Juan: "The Indians of Europe: the role of Spain's Enlightenment in the making of a global science". In: Astigarraga, Jesús (ed.): *The Spanish Enlightenment revisited*. Voltaire Foundation: Oxford 2015, pp. 83–103.

Pimentel, Juan: *Testigos del mundo. Ciencia, literatura y viajes en la Ilustración*. Marcial Pons Historia: Madrid 2003.

Ponz, Antonio: *Viaje de España*. 4 Volumes. Aguilar: Madrid 1988.

Prat, Dale / Gordon, Barbara: "The Environment and Nineteenth- and Twentieth-Century Spanish Literature". *Literature of Nature: An International Sourcebook*. Fitzroy Dearborn Publishers: Chicago 1998, pp. 248–256.

Rodríguez de Campomanes, Pedro: *Viajes por España y Portugal*. Editorial Miraguano: Madrid 2006.

Rueda, Ana: "Jovellanos en sus escritos íntimos: el paisaje y la emoción estética de «lo sublime»". *Revista de Literatura* LXVIII (136), July-December 2006, pp. 489–502.

Sitter, John: "Eighteenth-Century Ecological Poetry and Ecotheology". *Religion & Literature* 40 (1), Spring 2008, pp. 11–37.

Thion Soriano-Mollá, Dolores (ed.): *La naturaleza en la literatura española*. Academia del Hispanismo: Vigo 2011.

Urteaga, Luis: "Explotación y conservación de la naturaleza en el pensamiento ilustrado". *Geocrítica* 50, March 1984, pp. 7–46.

Natalia Álvarez Méndez
Subject and Landscape: Encounters with Nature in Contemporary Spanish Narrative

Abstract: As we recognize the evolutionary inflection points of the description of landscapes in our literary tradition, it is worthwhile to consider the trends in current Spanish literature. In the last decades, Spanish narrative has located plots in an indisputably urban environment. Nevertheless, there is a striking number of works that give special attention to the natural landscape, setting out the return to the rural as a response to the crisis, both economic and moral, of our society. The failure of modernity makes us aware of the necessity of reforging bonds with the natural landscape, communal space, and forgotten memory of a vital philosophy. Apart from alluding to renowned authors, such as Julio Llamazares, Luis Mateo Díez, and José María Merino, this study will focus its attention on several titles of a younger generation, such as *Belfondo* (2011) by Jenn Díaz, *Lobisón* (2012) by Ginés Sánchez, *Intemperie* (2013) by Jesús Carrasco, *El bosque es grande y profundo* (2013) by Manuel Darriba, *Por si se va la luz* (2013) by Lara Moreno, *El niño que robó el caballo de Atila* by Iván Repila, and *Alabanza* (2014) by Alberto Olmos, among many others. On the whole, these novels dissociate the rural from capitalist consumerism. They bring us closer to a return to the origins linked to an existentialism that highlights the prominent relations between man and earth. Additionally, they recover the agricultural landscape, recreate the life in small villages of our country, stress the adversities of the land, depict a hard and hostile surrounding, and denounce the ecological destruction of nature. From a necessary interdisciplinary perspective, which emphasizes ecocriticism among the current ways to study landscape, this analysis will focus on the reproach that is made in many of those narrations to the lost connection between man and earth. It will also analyze the ecological conscience perceptible in the aforementioned works that turn the environmental crisis into literary subject and suggest the necessity of recovering the harmonious coexistence between man and nature, of conceiving landscape as a biological being threatened by capitalism, by consumerism, and by the destructive industrialization of our societies.

Keywords: Ecocriticism, Landscape, Nature, Contemporary Spanish Narrative

Introduction

The thematization of space in literature can transform the landscape into a narrative element. Whether real or imagined, natural or artificial, the landscape may become a structural element—or even the main character—of the narrative, transcending its traditional role as setting or adornment and determining the development and meaning of the plot. As has been the case in painting with the rise of the pure or total landscape, this shift from background to foreground, from "parergon" to "ergon", has emerged as a textual phenomenon capable of sustaining the meaning of the work and the writer's worldview (Guillén 1996, pp. 68–69).[1]

Literary description of the landscape has undergone marked changes throughout history, affecting approaches to its study and interpretation. The defining moments in this evolution are well known in the Spanish literary tradition; nevertheless, it is of interest to examine the case of contemporary narrative. With rare exceptions, Spanish narrative since the 1980s has been set in unquestionably urban environments. Various critical studies have employed theories that analyze space from social, political, economic, cultural, and intimate perspectives, or have adopted the perspective of the "spatial shift" identified in the social sciences and humanities, to explore literary images of artificial, urban landscapes and their significance. Many have linked urban landscapes to modern and postmodern ideas of space, associated with social fragmentation and loss of the sense of belonging to a place. However, few have reflected on the existence or absence of the natural landscape, untouched by human intervention and owing to philosophical and cultural beliefs, often perceived as a manifestation of the otherness, as an external physical environment indifferent to humanity: "no tendríamos paisaje si el hombre no se retirase decisivamente de él, si su protagonismo no cesara de ser visible, si no se privilegiase esa clase tan radical de otredad que en ciertas épocas se ha llamado, con mayúscula, la Naturaleza" (Guillén 1992, pp. 77–78)[2] ["we would not have landscape if man did not retreat decisively from it, if his protagonism did not cease to be visible, if he did not privilege that radical class of otherness that in certain eras has been called, with a capital letter, Nature"].[3]

It should also be borne in mind that as a cultural construction, the landscape is intimately linked to the individual and to his or her interpretation:

---

1 "El hombre invisible. Paisaje y literatura en el siglo XIX". In: Villanueva, Darío y Cabo Aseguinolaza (eds.): *Paisaje, juego y multilingüismo. Actas del X Simposio de la Sociedad Española de Literatura General y Comparada*. Servicio de Publicaciones de la Universidad de Santiago de Compostela: Santiago de Compostela 1996, pp. 67–83.
2 Guillén, Claudio: "Paisaje y literatura: o, los fantasmas de la otredad". In: Vilanova, Antonio (ed.): *Actas del X Congreso de la Asociación de Hispanistas*, vol. I. Promociones y Publicaciones Universitarias: Barcelona 1992, pp. 77–98.
3 All translations by Ellen Skowronski-Polito.

Subject and Landscape 115

El paisaje no es, por lo tanto, lo que está ahí, ante nosotros, es un concepto inventado o, mejor dicho, una construcción cultural. El paisaje no es un mero lugar físico, sino el conjunto de una serie de ideas, sensaciones y sentimientos que elaboramos a partir del lugar y sus elementos constituyentes. La palabra paisaje, con una letra más que paraje, reclama también algo más: reclama una interpretación, la búsqueda de un carácter y la presencia de una emotividad (Maderuelo 2013, p. 38).[4]

[Landscape is not, therefore, what is there, before us; it is an invented concept, or better stated, a cultural construction. Landscape is not a mere physical place, but the set of a series of ideas, sensations and feelings that we elaborate from the starting point of place and its constituent elements. The word for landscape/*paisaje*, with one letter more than the word for place/*paraje*, also demands something else: it demands an interpretation, the search for a character and the presence of an emotionality].

As Guillén has stated, "es precisamente la mirada humana lo que convierte cierto espacio en paisaje, consiguiendo que una porción de tierra adquiera por medio del arte calidad de signo de cultura, no aceptando lo natural en su estado bruto sino convirtiéndolo también en cultural" (1992, p. 78) ["it is precisely the human gaze that turns a determined space into a landscape, achieving that a piece of land acquires the quality of a cultural sign through art, not accepting the natural in its raw state but also converting it into a cultural one"]. Thus, "el paisaje *natural*, la bella naturaleza, es—ni más ni menos que el resto de los objetos que pueblan nuestra conciencia—una creación *artística* del hombre" (Ayala 1996: 30)[5] ["the natural landscape, beautiful nature, is—neither more nor less than the other objects that populate our conscience—an *artistic* creation of man"]. Consequently, it is the subject's vision that endows the landscape with meaning, the subject who both produces and consumes physical space through the senses, which attribute emotional and aesthetic values to light, color, sound, smell and temperature. This process of subjectively inhabiting and experiencing our surroundings generates psychological spaces, or interior landscapes. Existential contemplation elevates these to the status of aesthetic objects that arouse a range of emotional sensations in the observer, as postulated in humanistic geography, giving rise to interactions between the landscape and the emotional states of the onlooker.

It would be difficult to cover all aspects of recent developments or conduct an exhaustive analysis of contemporary Spanish narrative; nevertheless, the foregoing

---

4 Maderuelo, Javier: *El paisaje. Génesis de un concepto*. Abada Editores: Madrid 2013.
5 Ayala, Francisco: "El paisaje y la invención de la realidad". En: Villanueva, Darío y Cabo Aseguinolaza (eds.): *Paisaje, juego y multilingüismo. Actas del X Simposio de la Sociedad Española de Literatura General y Comparada*. Servicio de Publicaciones de la Universidad de Santiago de Compostela: Santiago de Compostela 1996, pp. 23–30.

indicates the value of posing the following questions and interrogating small pockets of the literature to help evaluate the impact of landscape on this corpus, as well as the applicability of specific theories, including ecocriticism. Hence, to what extent does the natural world appear in recent Spanish narrative? What places are described and how are they depicted? Do these reflect our ethical and aesthetic concerns? What is the outcome of the encounter between the subject and landscape in fiction? Has the natural landscape regained transcendence? Has it become a positive setting in the face of conflicts experienced by the characters? Is it depicted as an otherness that conceals meanings and values that explain our world and our sense of belonging in it? A further goal will be to determine whether a schism persists between humanity and the natural environment.

Nature in Late Twentieth and Early Twenty-First Century Narrative

In contemporary Spanish narrative, the natural landscape has played a minor role in comparison with other settings. Some of the works in which landscape is the protagonist, without eschewing a critical examination of the contemporary world, tend towards writing the self, unease in the present, and experiences of grief and crisis. Numerous writers in the late twentieth and early twenty-first century have addressed the return to this setting in narrative. In an interesting study entitled "Para una reeducación de los sentidos: Naturaleza y ficción en el siglo XXI", ["For the Reeducation of the Senses: Nature and Fiction in the Twenty First Century"] Álvarez-Blanco examines the dialogue between subject and landscape in novels by Julio Llamazares, Jesús Moncada, Miguel Delibes, Luis Mateo Díez, Clemente Alonso Crespo, José Ángel González Sainz, Ramón Acín, Xurxo Borrazás, Isabel Cobo, and Juan Pedro Aparicio, focusing on the characters' encounter with the natural world as the result of a journey prompted by an experience of mourning, in the quest to overcome a state of existential crisis (p. 39).[6]

Álvarez Blanco's analysis of these novels allows her to reflect on "la importancia de una labor narrativa que escrita en un momento de vacío epistemológico propone en la reconexión del sujeto con el espacio un antídoto contra el automatismo propio de una era repoblada del espacio del simulacro"

---

6   Álvarez Blanco, Palmar: "Para una reeducación de los sentidos: naturaleza y ficción en el siglo XXI". In: Celma, María Pilar y González, José Ramón (eds.): *Lugares de ficción. La construcción del espacio en la narrativa actual*. Cátedra Miguel Delibes: Valladolid 2010, pp. 39–53.

(p. 40) ["the importance of a narrative work that, written in a moment of epistemological emptiness, proposes, in the reconnection of the subject with the space, an antidote against the automatism belonging to an era repopulated from the space of the simulacrum"]. Of particular interest is her reading of characters who see nature as mythical and inherited, perceptively explained by Anderson (1991) and Tönnies (2001). Both "señalan la existencia de un principio organicista sobre el que se funda la comunidad occidental y que considera que en la experiencia de la naturaleza, el sujeto descubre un tranquilizador sentido de pertenencia a un todo" (p. 40) ["point out the existence of an organicist principle on which the Western community is based and which considers that, in the experience of nature, the subject discovers a reassuring sense of belonging to a whole"]. Thus, longing for a sense of belonging in a communal space where humanity and the land are connected and perceiving the natural world from the standpoint of metaphysical organicism or transcendent pantheism, the characters find refuge through observation of the landscape, as well as the wisdom to face the anguish of solitude and individual destiny.

In some cases, an analysis of the impact of nature on these narratives demands a new avenue of study beyond aesthetic considerations of the landscape and the postulates of humanistic geography and cultural theory. In the novels by Llamazares and Díez, for example, the landscape is unquestionably identified as a cultural construct, a repository of collective memory, and a reflection and complement of the individual, who must not lose his or her sense of belonging to the place he or she observes and inhabits because it forms part of his or her identity. Thus, they champion rural cultures in the process of extinction and criticize industrialization for divorcing humanity from nature, with the inevitable social and spiritual consequences this entails. At the same time, other works demand an increased use of the postulates of ecocriticism.

In this respect, the focus, the gaze, is fundamental to the relationship between subject and object: "Cada forma de ver la tierra, cada manera de describirla o representarla supone que tras ella hay un tipo diferente de pensamiento" (Maderuelo, pp. 36–37) ["Every way of seeing the earth, every way of describing it or representing it supposes that behind it there is a different kind of thinking"]. In response, some new theories have adopted an interdisciplinary approach aimed at exploring the ontological background of the landscape in humanity's understanding of it. In recent decades, the environmental crisis has emerged as a theme in contemporary literature, awakening an ecological vision of the landscape that "puede servir tanto de nutritivo abono para la creatividad literaria como de renovadora perspectiva para la especulación teórica, el comentario crítico y la investigación histórica" (Marrero Henríquez 2006, p. 12) ["can serve both as a nutritious fertilizer for literary creativity and as a refreshing perspective for theoretical speculation, critical commentary and historical research"]. This turning point in landscape description is embodied in the awareness that humanity and nature are not separate, and the understanding that humans form part of the latter.

José Maria Merino is among the writers who espouse this vision of the natural environment. In some of his novels, the landscape assumes a leading role, with symbolic meanings and unique relationships with the subject it surrounds. On the one hand, he highlights the confrontation between nature and artifice. With diverse nuances in each case, he reflects on how the loss of a sense of belonging to a place—triggered by social fragmentation in the modern city—destroys the idea of community and condemns humanity to individualism. On the other hand, he explores the growing distance between subject and landscape, revealing the superiority of nature, derived from its savage force and its immutability as an element not subject to the transience that marks human life.

Notable among his works is his trilogy *Los espacios naturales*, comprising the novels *El lugar sin culpa* (2007), *La sima* (2009), and *El río del Edén* (2012). The spaces in these novels host the natural world, presenting topographically and historically recognizable settings depicted as cultural constructs—*locus amoenus*, utopian, Arcadian places transformed into chimeras and lost paradises—and as psychological, interior landscapes that highlight the liberated consciousness of their observers. Located in crisis and engaged in journeys of escape or return, his characters recapture their memories through contemplation and being in nature, attempting to unravel its mysteries and finding the strength to face trauma and grief in its symbols and values. Subject and landscape are at the heart of the plot, because observation of the exterior leads to understanding of the interior and therefore, to self-knowledge, rediscovery and acceptance not only of the past, the present and reality, but also of the world and the characteristics of humans and nature. The schism between humanity and the natural landscape evidences the need for a sense of belonging in a communal space, for a project that involves both in a yearned-for union.

In this connection between subject and natural environment, the author's ecological consciousness cannot be ignored. An ecological reader must "enfrentar la separación que la estética romántica de lo sublime estableció entre lo humano y lo no humano y situar la otredad de la naturaleza fuera de la jerarquía que la somete al sujeto perceptor" (Marrero Henríquez 2004, p. 28) ["face the separation that the romantic aesthetics of the sublime established between the human and nonhuman and place the otherness of nature outside the hierarchy that submits it to the perceiving subject"]. This stance is reflected in Merino's work, which presents a moral judgement on our social and individual failures, on humankind and its reality, characterized by the loss of values and ideals in consequence, among many other reasons, of an erosion of the sense of belonging due to the absence of any connection with the land. His work indicates the need to restore harmony between humanity and nature, to stop building hostile cities and to perceive the landscape as a living being menaced by society's capitalism, consumerism and destructive industrialization. The novels that make up his trilogy contain constant references to humanity's abuse of the natural environment.

This theme is intensified in some of his other works. Merino can thus be seen as writing within the genre termed eco-literature, which, as defined by Paredes and McLean, presents discourses that call for a reexamination of our relationship with the environment, urging us to respect and protect it rather than exercising privilege over it (pp. 6–7).[7] This perspective is fully developed in one of his short story collections, *Las puertas de lo posible (Cuentos de pasado mañana)*, set in a post-apocalyptic future marked by technological progress, totalitarianism, and human alienation.[8] Nature is utterly exploited, all forms of life and resources are devastated, and humans are controlled by technology, approaching the concept of the postmodern sublime proposed by Hitt in reference to undeniable worldwide environmental destruction.[9] In opposition to the artificial and oppressive space of the city, nature offers the possibility of liberation through observation as a means to reconnect humankind with the land. Once again, the sense of belonging to a place is portrayed as the solution to present and future loneliness and individualism. Restoring a dialogue with nature from a non-anthropocentric perspective is the only means to counter a virtually constructed identity and combat alienation from a space and its history in a world with no place for humans in a postmodern simulacrum of roots.

Subject and Landscape in the New Spanish Narrative

In this the second decade of the twenty-first century, a new literary movement termed neoruralism has emerged in response to the economic crisis and city dwelling. Rural flight and the depopulation of the Spanish countryside in favour of the cities—portrayed by Sergio del Molino in *La España vacía. Viaje por un*

---

7   Paredes, Jorge / McLean, Benjamín: "Hacia una tipología de la literatura ecologista en el mundo hispano". *Ixquit* 2, 2000, pp. 1–37.
8   *Las puertas de lo posible. Cuentos de pasado mañana*. Páginas de Espuma: Madrid 2008.
9   Hitt, Christopher: "Toward an Ecological Sublime". *New Literary History* 30.3, 1999, pp. 603–623.

País que nunca fue (2016)[10]—highlights the failure of modernity, as explored by John Berger in *Puerca tierra* (2001).[11] However, the number of people returning from the city began to rise following the eruption of the crisis in 2008 coupled with a growing awareness of the need to reconnect with the countryside, with communal space, and with the neglected memory of a life philosophy.

In recent years, the twenty-first century neorural novel has explored this situation, examining society's economic and moral crisis and advocating a return to nature. This period has witnessed the emergence of a young generation of writers aware of the reality of village life. Familiar both with post-war rural literature, written by authors such as Cela, Delibes, Martín Gaite, and Benet, and later works such as those by Llamazares and Díez, these authors are producing remarkably poetic narratives. Notable examples of these include *Belfondo* (2011) by Jenn Díaz, *Lobisón* (2012) by Ginés Sánchez, *Intemperie* (2013) by Jesús Carrasco, *El niño que robó el caballo de Atila* (2013) by Iván Repila, *El bosque es grande y profundo* (2013) by Manuel Darriba, *Por si se va la luz* (2013) by Lara Moreno, *Las efímeras* (2015) by Pilar Adón, and *Meteoro* (2015) by Mireya Hernández.

Not always set in identifiable spaces or times, these novels contrast rural life and capitalist consumerism. In some cases, they depict a return to one's roots from an existentialist perspective that exalts the centrality of humankind's relationship with the land. Portraying life in Spain's small, remote, and depopulated villages, they also recapture agricultural landscapes and vocabulary, allude to legends, and depict the harshness of the land and the brutality of survival and pain in this environment. The images and values associated with the natural landscape are largely conditioned by the opposition between rural and urban life. Some of the novels spotlight the different emotional ties that locals and outsiders form with the countryside, through the classic encounter in which a villager connects with or confronts a city dweller with ancestral wisdom. Others focus on the construction of symbolic, mythical and allegorical spaces. Many of them highlight the brutality of the land, portraying a harsh, hostile environment in line with Virgil's *Georgics* and closer to a negative rather than a bucolic view of rural life. The bleakness of the rural environment is accompanied by stark relationships of dependency between people that lead to resentment, possibly severe violence, and events that place the characters in extreme situations.

Use of the term neoruralism, whether apt or not, does not imply that the abovementioned authors all share exactly the same aesthetic. The landscape, the

---

10 Molino, Sergio del: *La España vacía. Viaje por un país que nunca fue.* Turner: Madrid 2016.
11 Berger, John: *Puerca tierra*. Punto de lectura: Madrid 2001.

natural world and rural life all appear in their works, but the rich existential or dystopian nuances in their narratives, combined with realist or even mimetic elements, cannot be reduced to a genre that refers exclusively to rural narrative. Where they all coincide, however, is in their departure from the predominantly urban literature of contemporary times by taking the present-day rural world and the situation of Spanish villages as their literary theme and exploring the human condition through the encounter between humankind and nature. The unquestionably critical gaze underlying these existential plots presents us with the abandonment of the rural world and delves into the consequences of this and the relationship between humans and the natural landscape.

Claustrophobic rural environments proliferate in the novels by these authors, for example in *Belfondo* by Jenn Díaz, whose novel is set in a remote Spanish village, and *Lobisón* by Ginés Sánchez, which is set in a rural world that alludes to legends.[12] However, it is *Intemperie* by Jesús Carrasco that many consider the novel that, with echoes of Delibes, ushered in neoruralism in Spanish literature.[13] Hostile nature assumes the role of a character in this book, in which the main protagonist, born into this environment and endowed with rural wisdom, fights a desperate struggle against the violence of the local men and against a landscape that conveys desolation and torment and encompasses a rural perspective despite being perceived artistically. The scorched earth, the remains of furrows and fields, and the infinite, arid plain delineate a closed world at an unspecified time in which the main character must strive against hunger and pain before discovering true human dignity.

Iván Repila presents us with a different vision of the landscape in *El niño que robó el caballo de Atila*.[14] Writing in the style of a folktale, the author creates a fascinating allegory of fighting spirit in an attempt to inspire an entire generation to rebellion, criticism, and solidarity. This is clearly evidenced by the two quotes that introduce the novel: the first—a statement by Margaret Thatcher concerning free trade and the free market—contradicts the view of neoliberalism as increasing the gap between rich and poor countries, while the second— the words of Bertolt Brecht—presents an urban context of hunger and disorder in which the only option is to rebel. The desire for change, the desire to construct a new era, is conveyed in the passages relating dreams, which portray the hunger of the many trapped below, in the pit, compared with the few that are

---

12 Díaz, Jenn: *Belfondo*. Principal de los Libros: Barcelona 2011; Sánchez, Ginés: *Lobisón*. Tusquets: Barcelona 2012.
13 Carrasco, Jesús: *Intemperie*. Seix Barral: Barcelona 2013.
14 Repila, Iván: *El niño que robó el caballo de Atila*. Libros del Silencio: Barcelona 2013.

above, where power is exercised. It is also evident in the allusion to cities in which people are subjugated and bewildered, and with the hope that all this will culminate in uprising and overthrow.

Again set in an unspecified time and place, this political and economic reading of the injustice, inequality, and oppression that plague contemporary societies is based on an intense and raw account of the human condition, on the symbolic struggle for survival of two brothers, Grande, who is entirely pragmatic, and Pequeño, who is an absolute dreamer. Both are trapped in a seven-meter deep pit in the heart of a menacing forest. Their struggle to escape culminates in the sacrifice of Grande and in the revenge sworn by Pequeño and carried out against their own mother, who blithely condemned them there to die. Ants, green snails, yellow worms, soft roots, tiny larvae and muddy water comprise their only sustenance and do not allay nausea, weakness, exhaustion, delusions, and hallucinations.

As in *Intemperie*, the brothers in Repila's novel have an intimate knowledge of the land and its moods because they have grown up beneath its sky, enabling the author to portray the adversities of rural life and show that the landscape forms the identity of the people who inhabit it:

> Saben que en este mes un sol feroz significa la llegada inminente de una tromba de agua. Lloverá porque siempre llueve cuando la carne se despelleja, y porque en estos campos parece gobernar una mecánica del sufrimiento, según la cual a toda decisión de la naturaleza le replica su contraria. Por este motivo las personas son, aquí, severas de piel y de carácter, y enfrentan las sanciones que les impone la tierra con rigurosa paciencia, sin demandas ni quejas, aunque ello les suponga una fractura en la comunicación emocional, en el contrato humano de la convivencia y en la gestión de los afectos. Los hermanos son prueba de ello. Han dejado de mirarse a los ojos, de buscarse en el otro como hacían los primeros días. Las muestras de cariño son innecesarias cuando rige la conservación. El amor como un pacto de silencio donde se administran violencias propias de un reptil, de un cocodrilo viejo (pp. 37–38).

> [They know that in this month a fierce sun means the imminent arrival of heavy downpours. It will rain because it always rains when the flesh peels, and because in these fields a mechanism of suffering seems to govern, according to which every decision of nature replicates its opposite. For this reason the people here are severe of skin and character, and they face the sanctions imposed by the land with rigorous patience, without demands or complaints, even if this supposes a fracture in emotional communication, in the human contract of coexistence and in the administration of affections. The brothers are proof of that. They have stopped looking into each other's eyes, looking at each other as they did the first days. The examples of affection are unnecessary when conservation governs. Love as a pact of silence in which cruelties are dispensed befitting a reptile, an old crocodile].

Meanwhile, *El bosque es grande y profundo* by Manuel Darriba is immersed in an apocalyptic aesthetic.[15] The statement on the back cover of this novel, "En el principio fue el verbo, es decir la ficción, es decir, el Ogro" ["In the beginning was the word, that is to say fiction, that is to say, the Ogre"], suggests an association with the meaning that the forest has in children's stories. As in the novel by Repila, it also deploys the motif of siblings abandoned in a forest, linking the loss of innocence to the experience of harm, death, and the illusory nature of paradise. This effect is underlined by the fact that the main characters—and parts of the novel—are called Hansel and Gretel. In a dystopia caused by the outbreak of the final war, people have abandoned the city and fled through a gorge to the forest. The city is threatened by attacks, dust, groups of molecules, clouds of atoms, flames, and ashes, but the natural world offers no shelter either.

Once again, the harshness of rural work and subsistence is described with Georgic overtones. Some of the descriptions of the surrounding landscape read like oil paintings, pictures, or mosaics, revealing an artistic perspective in the gaze of the narrator. However, these poetic brushstrokes, which sometimes depict the beauty of nature, also evidence its otherness, its configuration as an element divorced from humankind. In this natural environment, in which the seasons pass as food supplies dwindle and the savage, cruel war runs its course, a shocking social regression occurs. The human condition is ever more strongly linked to primitive instincts, the struggle for survival, madness, and brutalization. There may be nothing left in the city, but the forest will not harbor civilization either; rather, it represents a menacing, inhospitable space in which all moral sense has been lost. The forest is large and deep, harsh, alive, and full of anger: "—El bosque es grande. Da miedo. /—Es lo que hay. Es la vida. /—Es la vida. Da miedo" (p. 45) ["—The forest is big. It's scary /—It is what it is. That's life. /—That's life. It's scary"].

The brutality of survival in the rural world, in depopulated natural settings, is once again the theme of other novels where symbolic and mythical spaces cede their protagonism to interesting but dystopian worlds created in response to the current economic crisis. Despite what one might expect, some of the most notable of these do not offer a reading of otherness based on ecological values that eulogize the rural compared with the urban world, but instead depict the differing bonds formed by locals and outsiders with the land, its inhospitable nature, and the bleak relations between its inhabitants. Two of these worlds are described in *Por si se va la luz* by Lara Moreno and *Las efímeras* by Pilar Adón

---

15 Darriba, Manuel: *El bosque es grande y profundo*. Caballo de Troya: Barcelona 2013.

in both of which "el ecologismo y su idealismo de la naturaleza es deconstruido" (Pozuelo Yvancos, p. 362)[16] ["the ecologism and its idealism of nature are deconstructed"]. Moreno and Adón have adopted a dystopian approach to the crisis novel, locating "de manera dialéctica la relación entre los personajes y el entorno, creando un contexto propio que es alienado, enajenado respecto a la referencia realista, pero con un valor crítico no menor" (Pozuelo Yvancos, p. 352) ["the relationship dialectically between the characters and the environment, creating a context of its own that is alienated, distanced from the realistic reference, but with no less of a critical value"].

We should not forget the prescient capacity of this genre, which fosters critical speculation on the present. In *Por si se va la luz*, their flight from civilization immerses the novel's characters in a depopulated, natural realm that serves as a critique of contempt for the land and the collapse of the relationship between humankind and nature, revealing the adversities and harsh struggle for survival in this physical environment. The novel is divided into two large sections entitled "Invierno" and "Verano", underlining the climatic extremes that mark a suffocating atmosphere from the outset. The use of multiple perspectives, in which the male and female voices of various characters alternate in third person, hinders the reader from forming a full picture of the plot. For example, the background and future of the characters remain unknown, as do many aspects that affect their present throughout the novel; however, these gaps contribute to the richness of the novel, which grips the reader with its intensity and moving subject matter.

The narrative opens with the voices of Martín and Nadia, a young couple who have left the city for a village, an act that exposes them to extreme situations. Their neighbors comprise Enrique, who had made the same life change many years earlier, and two elderly locals, Damián and Elena. These are later joined by Ivana and Zhenia. The reason for this move from the city to the country is dystopian: the narrative speaks of a danger that is causing suffering and will destroy humanity, although the precise nature of this threat is never clarified. The reader must act as an active accomplice, filling in the gaps in the information. It is not difficult to detect the current crisis in this critique of a system in which politics cannot offer a solution to the coming apocalypse because it is governed by an absurd, insane economy. The organization that helps the protagonists to move is also under suspicion: it is not known "a qué engranaje responde" (p. 102) ["to which gear it responds"], and there is a suggestion that it may be watching

---

16 Pozuelo Yvancos, José María: *Novela española del siglo XXI*. Cátedra: Madrid 2017.

the characters.[17] Furthermore, Damián is obsessed with the apocalypse, which will occur "cuando todo el mar se abalance y decida envolver otra vez la Tierra entera" (p. 128) ["when the entire sea surges forth and decides to envelop the whole Earth again"] while sporadic allusions are made to a pandemic, associated with the appearance of terrifying dogs, and animals and people who fall ill and die because of a virus, or perhaps as a result of "la mala alimentación, la suciedad, el caos, el líquido contaminante que se vertía de las ciudades hacia el campo" (p. 248) ["Poor nutrition, filthiness, chaos, the polluting liquid that was poured from the cities into the countryside"].

The portrayal of the city is clearly negative. It is described as an asphyxiating space where young people have no future, despite their training; as an oppressive, alienating place from which flight into the countryside is the only possible option for the protagonists. Nonetheless, their attitudes towards this decision and its consequences differ. Nadia knows that she carries within her the fumes that she has inhaled there. She is also aware of environmentally unfriendly and exploitative consumerism. She speaks, for example, of "mi ropa interior de fibra sintética que me provocaba picores o el armario de los cosméticos lleno de cremas fabricadas con esencia de placenta y quizá de fetos" (p. 116) ["my synthetic-fiber underwear that made me itchy or the cosmetics cabinet full of creams manufactured with placenta essence and perhaps fetuses"]. Mention is also made of carcinogenic food, relay masts, destructive particles of plastic (p. 161) and the problem of poorly organized recycling (pp. 240–241). Neither is humankind's subjection to technology omitted: "Salíamos del más absoluto de los descontroles, que es la dependencia virtual, el monstruo que se fagocita a sí mismo" (p. 103) ["We left the most absolute of the mayhems, which is virtual dependency, the monster that devours itself"]. Meanwhile, Martín observes that gardens and squares with trees have gradually disappeared from the cities, swallowed up by irresponsible and unsympathetic development:

> Luego el meollo. La absorción y el derrame. Lo monstruoso. La frivolidad del ensanchamiento, esos kilómetros llenos de construcciones, de pequeñas ciudades que nunca terminaron de existir, bloques simétricos con sus instalaciones de luz y de agua, urbanizaciones parásito. Hombres parásito. Virtualidad y desorden. Es curioso que virtual y virtud tengan la misma raíz. Ahí empieza el precipicio, la estafa (p. 101).

> [Then the core. The absorption and the spill. The monstrous. The frivolity of the widening, those kilometers full of construction, of small cities that never finished existing, symmetrical blocks with their installations of light and water, parasitic urbanizations.

---

17 Moreno, Lara: *Por si se va la luz*. Lumen: Barcelona 2013.

Parasitic men. Virtuality and disorder. It is curious that virtual and virtue have the same root. There begins the precipice, the scam].

The young couple is conscious of the finitude of the artificial compared with nature. Nadia reflects on the rapid deterioration of the city: "primero llegaron los recortes y luego las restricciones, el paraíso construido por el hombre siempre tiene un mal morir" ["first came the cuts and then the restrictions, the paradise built by man always has a bad death"], whereas in the countryside, "lo construido por el hombre se parece un poco más a lo construido por la naturaleza y su abandono puede llamarse ruina en vez de purgatorio" (p. 115) ["what is built by man looks a little more like what nature has built and its abandonment can be called ruin instead of purgatory"]. Martín gives a stark description of the progress achieved by a self-destructive system, referring to the thousands of dwellings in the world that have been abandoned, although they are still connected to the grid. Of these empty buildings, he says: "Quizá sirvan exactamente para eso, para meter gente dentro, gente que no será capaz de levantarse del cemento para cavar en el cemento. El mundo construido se convertirá en un gran campo de concentración" (p. 104) ["Maybe they will serve exactly for that, to put people inside, people who will not be able to get up from the cement to dig in the cement. The built world will become a great concentration camp"]. Despite what one might expect, the countryside is not lionized in comparison with the city, as the author "ha huido de paisajes idílicos y de una supuesta vida alternativa y percibe una historia alejada del idealismo ecologista" (Pozuelo Yvancos, p. 366) ["has fled from idyllic landscapes and a supposed alternative life and perceives a history far from ecological idealism"].

Nonetheless, their move reflects values of non-consumption, since their new environment requires that they renounce all material and technological goods, and supposedly promises liberation from stark relations of dependency and domination with family and friends, as a consequence of their departure from their world. These premises suggest a struggle between an idyllic and a dystopian vision, although in the brief moments when it seems possible to "confiar de nuevo en el inagotable sustento de la Tierra" (p. 84) ["trust again in the inexhaustible sustenance of the Earth"] and escape from their problems by connecting with and enjoying nature, the harshness of the environment intrudes. Martín, who sees the countryside as a beautiful place, a landscape worthy of being photographed or painted, is nevertheless conscious that "hay algo anormal en el ambiente" (p. 101) ["There is something abnormal in the environment"]. Nadia, on the other hand, would have preferred to endure in the city to the end, despite her awareness that it represents a world of artificial well-being (p. 17).

Life in nature does not imply a return to Arcadian lifestyles: quite the contrary, it appears to represent a regression. They have left behind the virtual dependence of modern city dwellers, but, as they recognize, "entramos en otro tipo de dependencia, más elemental, más cercana, pero que demuestra al fin y al cabo lo frágiles e inexpertos que somos" (p. 103) ["we enter into another type of dependence, more elementary, closer, but which shows in the end how fragile and inexperienced we are"]. Relationships with previously unknown people pose problems. In addition, their lives are threatened by violence, and they experience intense emotions and dissatisfactions that infiltrate their personalities. The countryside presents them with the problems of survival, lack of experience in agriculture and livestock, and adverse weather conditions. They only survive thanks to the help of their neighbors, who were born there or have lived there for a long time. The struggle for survival is set in a closed world overshadowed by difficulties and distrust.

The urban couple's estranged relationship with the land is clear, illustrated in their attempts to integrate in this environment while perceiving its otherness. Resistant to this arcane world whose wisdom she doubts, and wary of the need to play at communities in order to survive, Nadia offers the most negative vision. In reference to Martín, she muses: "Desde luego no quiere ver que nuestros pulmones ya nunca estarán llenos de aire puro ni que esta forma de existencia responde a los mismos patrones de hipocresía; mucho menos que el orden aquí establecido peligra con idéntica fragilidad" (p. 83) ["Of course he does not want to see that our lungs will never be full of pure air anymore or that this form of existence responds to the same hypocritical patterns; much less that the order established here is in danger of the same fragility"].

For his part, Martín is more convinced of the need to flee to this place, yet recognizes that the harmony and safety he had hoped to find there are as unattainable as in the city: "yo soy consciente de que solo nos hemos alejado; estando aquí, construyendo en esta aldea un modo de subsistencia, estamos retrasando lo inevitable. Los cimientos se convulsionan y el derrame alcanzará cualquier lugar" (p. 102) ["I am aware that we have only moved away; being here, building a mode of subsistence in this village, we are delaying the inevitable. The foundations convulse and the spill will reach anywhere"].

The remaining characters do not offer a consistently positive view of rural life either, since the countryside is not depicted as a protective space or refuge that offers its inhabitants relief from their inner concerns. Dragging her daughter Zhenia with her, Ivana embarks on an oppressive regression in this inhospitable environment from which neither will return. Living far from the modern world and the frenetic pace of life in the city, Enrique finds some relief from his distress, but he is not entirely free of it, and faced with the extreme situation related at the end of the novel, he dreads the uncertain fate of the village, now transformed into an island from which there is no escape. The pain provoked by family relationships is embodied in Elena and her tragic experience with her daughter. In consequence, she does not trust the young women and calls the girl Zhenia a demon. Surrounded by darkness in the closed, claustrophobic atmosphere of her unlit home, the old woman knows the harshness of life in the countryside and the secrets of rural wisdom. The scenes in which she appears are sometimes sordid—giving birth in the pigsty, the sick animals with which she lives—and portray the profound misery caused by mistakenly thinking that "los humanos pueden ser seres inofensivos" (p. 99) ["humans can be harmless beings"]. Born in the village, Damián is aware of the unstoppable force of nature: "La naturaleza no es lenta, es una vorágine. Un vendaval. Como el que llegará" (p. 46) ["Nature is not slow, it's a vortex. A gale. Like the one that will arrive"]. It is also a symbol of the characters' internal struggles, not only with "la Grande (la muerte)", "la víbora que crea el mundo y se lo lleva, no podemos engañarla", but also with "la Pequeña", "la culebrilla que nos come por dentro. Va ocupando cada vez más sitio en el pecho y nos acostumbramos a llevarla" (p. 127) ["The Great One (death)", "the viper that creates the world and takes it away, we cannot deceive it", but also with "the Small One", "the shingles that eat us from inside. It occupies more and more space in our chest and we get used to carrying it"]. Regardless, he believes that the countryside can be a liberating force for humanity, a place to rediscover one's true essence: "Uno se hace fuerte con lo fuerte [...] Imagino que el hombre más envilecido de todos es aquel al que no le importan la lluvia o las sequías" (p. 127) ["One becomes strong with the strong [...] I imagine that the most debased man of all is the one who does not care about rain or drought"].

In this dystopia where the light of the world seems to be fading, vanquished by a lack of food and a pandemic that is devastating the land, a theme emerges of societies created by economic and ideological forces and of harsh human relationships in a context where humans must struggle against nature. Despite constant distrust and the unpredictability of fleeing or starting anew, some of the characters find hope: "De alguna manera, en un punto delicado, la niña, la mujer joven y el viejo forman un trío sin edad que se alimenta de sí mismo. Es el milagro de algunos animales que conviven en paz" (p. 234) ["Somehow, at a delicate point, the girl, the young woman and the old man form an ageless trio that feeds on itself. It is the miracle of some animals that live in peace"]. As Pozuelo Yvancos has said, "Es una novela radical en su sentido etimológico, porque sitúa a sus personajes en una raíz donde lo primigenio humano termina rescatándolos de la ley de la naturaleza" (p. 367) ["It is a radical novel in its etymological sense, because it places its characters in a root where primal humanity ends up rescuing them from the law of nature"]. Pilar Adón employs a beguiling style full of suggestion, disquieting insinuation, ellipsis and powerful imagery to create an asphyxiating and strange world in Las efímeras.[18] The countryside figures largely, hosting a community that has fled civilization to live in an unidentified place. Once again, contact with the landscape does not engender a more authentic, enriched existence. The Arcadian life is replaced by a vision of the harshness of rural life, far removed from "toda idealización de la naturaleza, que convertida en un ser vivo es fuente de peligros y depredaciones" (Pozuelo Yvancos, p. 368) ["all idealization of nature, turned into a living being, is a source of danger and depredation"].

> A la cita de John Fowles, "El caos verde. O un bosque", que sugiere el desorden o el orden posible en el entorno, se suma la descripción del espacio narrativo, poblado de vegetación, que se percibe protector—las encinas nombradas por Dora Oliver, como en los cuentos infantiles, y la contemplación idílica de arboledas y senderos—, o amenazante—cuando los humanos invaden el territorio salvaje o se pierden en el monte. En este lugar natural indeterminado, rasgo reiterado en las obras de la autora, que permite universalizar las experiencias de los personajes, se sitúa la comunidad de la Ruche, evocadora de la escuela libertaria francesa de igual nombre creada a principios del siglo XX. Este nuevo microcosmos pretende ser perfecto, pero la utopía se desmorona pronto. Las emociones y los afectos humanos son origen de la destrucción. (Encinar, pp. 19-20)[19]

> [Regarding the quote by John Fowles, "Green chaos. Or a forest", which suggests the disorder or possible order in the environment, adds the description of the narrative space, populated by vegetation, perceived as protective—the oaks named by Dora Oliver, as in children's stories, and idyllic contemplation of groves and paths—, or threatening— as when humans invade wild territory or get lost on the mountain. In this indeterminate natural place, a feature reiterated in the author's works, which makes it possible to universalize the characters' experiences, resides in the community of la Ruche, reminiscent of the French libertarian school of the same name created at the beginning of the twentieth century. This new microcosm pretends to be perfect, but the utopia soon crumbles. Human emotions and affections are the origin of destruction].

In these apparently utopian microcosms, the restoration of community life and the rejection of everything unnecessary and superfluous, so as not to strain its bounty, does not bring tranquillity for its members, represented in this novel by the siblings Oliver, Denise, Anita, and Tom. Community is not restored; instead, conflict is avoided through indifference and lack of interest in the others.

---

18 Adón, Pilar: *Las efímeras*. Galaxia Gutenberg: Barcelona 2015.
19 Encinar, Ángeles: "Dominio, sumisión y dependencia: motivos recurrentes en las obras de Pilar Adón y Sara Mesa". *Ínsula* 835–836, 2016, pp. 19–22.

Relations of power and submission emerge between the characters, unleashing repressed hatreds that shatter the apparent harmony with unbridled violence and brutality: "Y, por supuesto, todos ellos sabían matar. No había nadie en la comunidad que no pudiera hacerlo o que no hubiera visto cómo se hacía. Convivían con animales muertos, desollados, con heridas abiertas y gemidos de dolor" (Adón, p. 113) ["And, of course, they all knew how to kill. There was no one in the community who could not do it or who had not seen how it was done. They lived with dead animals, skinned, with open wounds and moans of pain"]. It is no coincidence that the second quote introducing the novel, from Marcel Proust, is: "No son hombres, son leones" ["They are not men, they are lions"]. The despotic, tyrannical nature of family and couple relationships intensifies the resentments felt by the members of this strange, sometimes sectarian, society that seems to function as a hive. An artificial order prevails, controlling the atmosphere in an attempt to render it inoffensive but actually heightening violence and domination.

The landscape is sometimes depicted from an artistic perspective, focusing on smells, colors, and sounds. But rural life offers an image of "un mundo hostil, nada halagüeño, gobernado por elementales fuerzas que se imponen a las criaturas y de las que ellas mismas son consecuencia" (Pozuelo Yvancos, p. 368) ["a hostile world, bleak, governed by elemental forces that impose themselves on the creatures and of which they themselves are a consequence"]. Prominence is given to the harshness of working on the land and the constant effort this requires, in an environment that nature may reclaim at any moment and where humanity is alien. This situation is confirmed by the brutalization which appears with growing frequency in the descriptions of the inhabitants, associating them with an increasing release of primitive instincts and savagery. The novel highlights the otherness of nature, its immeasurable force, capable of resisting the threat posed by human exploitation and indifferent to human events. It explores the failure of utopias caused by the "perversión de las relaciones entre los seres humanos. Frente a las pasiones y a los conflictos de los individuos permanece incólume la naturaleza, su fuerza y su poder resplandece y sólo ella proporciona la quietud" (Encinar, p. 20) ["perversion of relationships between human beings. In the face of the passions and conflicts of the individuals, nature remains unaffected, its strength and power shine and only it provides the stillness"]. Other novels published by this group of authors examine similar concepts of the rural world and nature, focusing on humanity's relationship with the land. For example, *Es un decir* (2014) by Jenn Díaz, *Meteoro* (2015) by Mireya Hernández, and *La tierra que pisamos* (2016) by Jesús Carrasco, among others, depict life in villages overshadowed by hostile atmospheres, even when the characters have fled there to escape the economic crisis, and they portray landscapes that are not bucolic, once again highlighting the problems derived from our rural past.

Conclusions

An analysis of these novels confirms the need to adopt an interdisciplinary approach in order to fully comprehend the realist, symbolic, and allegorical intricacies of these fictional responses to the crisis. Ecocriticism constitutes an essential part of studies of the landscape, shedding light on the critique of humanity's lost connection with the land inherent in many of these novels. Thus, the landscape is not viewed solely as a literary motif, as occurs with stylistic theory, and the postulates of cultural theory are enriched, since the land is seen as a concept, enabling a deeper understanding of its possible significance. The rural world is presented in these novels as a cultural element associated with identity, the community, and its history. To varying extents, the novels indirectly allude to collective memory and contain the traces of various historical episodes that, although indeterminate and imaginary, provoke reflection on the lack of moral progress in contemporary society and the illusory nature of utopias.

The desire to put an end to the appropriation and domination of nature underlines the pertinence of recent ecocritical studies of landscape and its associated values, which "se han redescubierto en estos últimos años por vías muy diferentes en un abanico que se abre desde el diletantismo artístico hasta el activismo ecologista, pasando por la práctica urbanística, las actividades turísticas o el positivismo biológico" (Maderuelo, p. 37) ["have been rediscovered in recent years by very different tracks ranging from artistic dilettantism to environmental activism, reaching urban practice, tourism activities or biological positivism"]. Ecocriticism sheds light on the ecological consciousness underlying some of these novels, which transform the environmental crisis into literary subject matter. Many of them suggest the difficulty of restoring harmony between humankind and nature, and depict the landscape as a living being menaced by society's capitalism, consumerism and destructive industrialization, all of which have heightened otherness and the threat of dystopia.

Bibliography

Adón, Pilar: *Las efímeras*. Galaxia Gutenberg: Barcelona 2015.

Álvarez Blanco, Palmar: "Para una reeducación de los sentidos: naturaleza y ficción en el siglo XXI". In: Celma, María Pilar y González, José Ramón (eds.):

Lugares de ficción. La construcción del espacio en la narrativa actual. Cátedra Miguel Delibes: Valladolid 2010, pp. 39–53.

Ayala, Francisco: "El paisaje y la invención de la realidad". In: Villanueva, Darío y Cabo Aseguinolaza (eds.): *Paisaje, juego y multilingüismo. Actas del X Simposio de la Sociedad Española de Literatura General y Comparada.* Servicio de Publicaciones de la Universidad de Santiago de Compostela: Santiago de Compostela 1996, pp. 23–30.

Anderson, Benedict: *Imagined Communities: Reflections on the Origin and Spread of Nationalism.* Verso: New York 1991.

Berger, John: *Puerca tierra.* Punto de lectura: Madrid 2001.

Carrasco, Jesús: *Intemperie.* Seix Barral: Barcelona 2013.

Carrasco, Jesús: *La tierra que pisamos.* Seix Barral: Barcelona 2016.

Darriba, Manuel: *El bosque es grande y profundo.* Caballo de Troya: Barcelona 2013.

Encinar, Ángeles: "Dominio, sumisión y dependencia: motivos recurrentes en las obras de Pilar Adón y Sara Mesa". *Ínsula* 835–836, julio-agosto 2016, pp. 19–22.

Díaz, Jenn: *Belfondo.* Principal de los Libros: Barcelona 2011.

Díaz, Jenn: *Es un decir.* Lumen: Barcelona 2014.

Tönnies, Ferdinand: *Community and Civil Society.* Cambridge UP: Cambridge-New York 2001.

Guillén, Claudio: "Paisaje y literatura: o, los fantasmas de la otredad". In: Vilanova, Antonio (ed.): *Actas del X Congreso de la Asociación de Hispanistas,* vol. I. Promociones y Publicaciones Universitarias: Barcelona 1992, pp. 77–98.

Guillén, Claudio: "El hombre invisible. Paisaje y literatura en el siglo XIX". In: Villanueva, Darío y Cabo Aseguinolaza (eds.): *Paisaje, juego y multilingüismo. Actas del X Simposio de la Sociedad Española de Literatura General y Comparada.* Servicio de Publicaciones de la Universidad de Santiago de Compostela: Santiago de Compostela 1996, pp. 67–83.

Hernández, Mireya: *Meteoro.* Caballo de Troya: Barcelona 2015.

Hitt, Christopher: "Toward an Ecological Sublime". *New Literary History* 30.3, 1999, pp. 603–623.

Maderuelo, Javier: *El paisaje. Génesis de un concepto.* Abada Editores: Madrid 2013.

Marrero Henríquez, José Manuel: "Del turista texual al lector ecológico". In: Santa Ana, Mariano de (ed.): *Paisajes del placer, paisajes de la crisis.* Fundación César Manrique: Lanzarote 2004, pp. 15–38.

Marrero Henríquez, José Manuel: "Conciencia ecológica y paisaje literario". In: Marrero Henríquez, José Manuel (coord.): *Pasajes y paisajes: espacios de vida, espacios de cultura*. Servicio de Publicaciones de las Palmas de Gran Canaria-Gabinete Literario: Las Palmas de Gran Canaria 2006, pp. 9–28.

Merino, José María: *El lugar sin culpa. Los espacios naturales*. Alfaguara: Madrid 2007.

Merino, José María: *Las puertas de lo posible. Cuentos de pasado mañana*. Páginas de Espuma: Madrid 2008.

Merino, José María: *La sima*. Seix Barral: Barcelona 2009.

Merino, José María: *El río del Edén*. Alfaguara: Madrid 2012.

Molino, Sergio del: *La España vacía. Viaje por un país que nunca fue*. Turner: Madrid 2016.

Moreno, Lara: *Por si se va la luz*. Lumen: Barcelona 2013.

Paredes, Jorge y McLean, Benjamín: "Hacia una tipología de la literatura ecologista en el mundo hispano". Ixquit 2, 2000, pp. 1–37.

Pozuelo Yvancos, José María: *Novela española del siglo XXI*. Cátedra: Madrid 2017.

Repila, Iván: *El niño que robó el caballo de Atila*. Libros del Silencio: Barcelona 2013.

Sánchez, Ginés: *Lobisón*. Tusquets: Barcelona 2012.

Latin American Ecocriticism

Scott DeVries

The Quiroga Frame: Animal Studies and Spanish American Literature

Abstract: I will document the most salient examples from Spanish American literature in which animal-ethical representations anticipate many of the most pressing concerns from present debates in animal studies. I identify moments from the corpus that articulate long-standing positions such as a defense of animal rights or advocacy for liberationism; that engage in literary philosophical meditations concerning mind theory and animal sentience; and that anticipate current ideas from Critical Animal Studies, including the rejection of hierarchical differentiations between the categories human and nonhuman. I begin the chapter with an enumeration of seven analytical tasks for a literary approach that I denominate "fauna-criticism". These points include reevaluations of the canon, the exposition of animal ethical positions from literary texts, and the identification of literary representation of debates within both traditional and critical animal studies as well as the ways in which literary expression might resolve these debates. The result is that this kind of analysis emphasizes the reframing of literary history with particular attention granted to the animal ethical positions of poetry and fictional prose, both in texts that have been considered canonical as well as those that have long been neglected.

Keywords: Animal Ethics, Ecocriticism, Spanish American Literature, Horacio Quiroga

The fauna-criticism that I will employ examines nonhuman sentience, animal interiority, and other ethical issues such as the livestock and pet industries, zoos, hunting, and species extinction. I will undertake a fauna-critical analysis of one well-known and one unfamiliar text from each of the major Spanish American literary periods: nineteenth-century *modernismo* and Regional, *indigenista*, and contemporary narrative. These will include Argentine Domingo Faustino Sarmiento's *Facundo: Civilización y barbarie* (1845) and *El Moro* by Colombian José Marroquín (1897) from the nineteenth century. Examples to be considered from the *modernista* corpus are "Estival" from Nicaraguan Rubén Darío's *Azul* (1888) and *Diana cazadora* (1917) by Colombian Climaco Soto Borda; from Regional texts, short stories by Uruguayan Horacio Quiroga and *Su nombre era muerte* (1947) by Mexican Rafael Bernal; and from the *indigenista* tradition, Peruvian Ciro Alegría's *Los perros hambrientos* (1939) and Ramón Rubín's *El callado dolor de los Tzotziles* (1948). Finally, from more recent narrative, I will consider Chilean Luis Sepúlveda's bestseller *Un viejo que leía novelas de amor* (1989) and fellow Chilean Fernando Raga's *Los hijos de Gaia* (2005).

Uruguayan Horacio Quiroga's short story "Historia de Estilicón" features an ape tamed by the narrator to act almost human, a dicey arrangement certainly, for what might happen when wild animals and human beings share the same living space: as it turns out in this story, there are hints of bestiality and murder. But as in many other examples from Quiroga's work, the representation of the ape here explores the parameters of nonhuman sentience and even blurs the line between absolute categories like "human" and "animal". As a result, Quiroga's texts operate within the conceptual territory of philosophically ethical thought with regard to notions such as agency, identity, moral duty, and ontological status but from the textual posture of narrative fiction. In short, Quiroga's stories, from within the literary aesthetic tradition of early twentieth-century Spanish American literature, address many of the issues that are relevant to present dialogues within several distinct fields of animal studies concerning ethics and human/nonhuman subjectivity.

Peter Singer is well-known for his explicit inclusion of nonhuman interests in the utilitarian treatise *Animal Liberation* published in 1975. This position was an early philosophical expression of human moral duty toward animals, but it was anticipated by utilitarianism's founding theorist Jeremy Bentham nearly 200 years earlier in 1785. In a passage that appears as a footnote within *The Principles of Morals and Legislation,* the philosopher considers applying utilitarian calculations to nonhumans: "The question is not, Can they [nonhumans] *reason?* nor, Can they *talk?* but, Can they *suffer?*"[1] While Bentham never explicitly pursued this animal ethical bent within his philosophy, Singer applied utilitarian principles *rigorously* and *extensively* to the moral status of animals. His argument comes down to the idea that "our present treatment of animals is based on speciesism, that is, a bias or prejudice towards members of our own species"[2] and that "to avoid speciesism we must allow that beings which are similar in all relevant respects have a similar right to life".[3] Singer's utilitarian liberationist position that the interests of animals should necessarily be taken into account— to not do so should be considered "speciesism"—when moral calculations are

---

1 Bentham, Jeremy: *The Principles of Morals and Legislation.* Hafner: New York 1948, p. 311.
2 Singer, Peter: "Ethics, Animals and Nature". In: Li, Hon-Lam / Young, Anthony (eds.): *New Essays in Applied Ethics: Animal Rights, Personhood and the Ethics of Killing.* Palgrave: New York 2007, p. 29.
3 Singer, Peter: *Animal Liberation.* Avon Books: New York 1977, p. 20.

made for the maximization of pleasure and the minimization of pain is considered a key moment in the development of animal studies.

Nevertheless, utilitarianism as an ethical theory has had its share of detractors. Regan, for one, holds that the problem with Singer's ideas is that they do not represent an "abolitionist" position; that is, if the utilitarian calculations come out just so, some morally objectionable actions towards animals may be permitted. As an alternative, Regan outlines a philosophy of rights granted to all who qualify as "subjects-of-a-life"; that is, those whose "desires, beliefs, and feelings have a psychological unity".[4] Regan then affirms that "nonhuman animals who concern us are like us in that they, too, are subjects-of-a-life", that they therefore possess inherent value, and that "all those human beings *and* all those animal beings who possess inherent value share the equal right to respectful treatment" (p. 96). I will refer to Singer's liberationist and Regan's rights theory approaches to animal studies and other animal-protectionist or animal-welfare reformist positions as Traditional Animal Studies (TAS). But despite Regan's attempt to correct the perceived deficiencies in Singer, the reservation of rights to animals is a notion that has been criticized in turn. For example, Huggan and Tiffin expose what they describe as human-centered bias to argue that "the very idea of rights, especially the granting or extending of rights to others of all kinds, may itself be regarded as in essence anthropocentric, since it is only the dominant (human) group that is in the position to do so".[5] Indeed, the field of animal studies extends far beyond utilitarian liberationism or rights theory; the idea of *Critical Animal Studies* (CAS) suggests as much by its terminology: the idea is not to simply advance ethical theories, but to engage critically with animal studies more broadly.

Animal studies of both forms would consider ideas from antiquity, such as when Plutarch and Porphyry declined to eat animals for food because they believed it was inherently unjust to eat rational beings, and the Cynics went so far as to assert the superiority of animals over humans in certain cases. Both would condemn Descartes who, in the seventeenth century, infamously referred to animal suffering as nothing more than the accidents of automata, something like the "wheels and weights" of a clock,[6] a position that largely held sway until

---

4 Regan, Tom: *Animal Rights, Human Wrongs: An Introduction to Moral Philosophy*. Rowman and Littlefield: Lanham, MD 2003, p. 64.
5 Huggan, Graham / Tiffin, Helen: *Postcolonial Ecocriticism: Literature, Animals, Environment*. Routledge: London 2010, p. 19.
6 Descartes, René: *Discourse on the Method*. In: Adler, Mortimer J. (ed.): *XXXI, Great Books of the Western World*. Hutchens, Robert Maynard (series ed.). Encyclopaedia Britannica: Chicago 1952, p. 60.

Singer. And both would consider questions on nonhuman sentience, such as "what does animal consciousness seem like?" and "do animals maintain a sense of identity as humans do?" The responses to these questions have had, at different times in history, profound consequences for how humans construed their moral duty to beings beyond our species. In this regard, DeGrazia's summation is apt: "the path to the ethical treatment of animals runs through their minds".[7] But the CAS perspective would go further and question the validity of human subjectivity as the basis for the construction of ethical theories regarding animals. Cavalieri, for one, has the character Alexandra in a dialogue called "The Death of the Animal" claim that "perfectionism—the hierarchical arrangement of the moral status of individuals based on (the level of) possession of certain cognitive skills—is an atavism that a sound ethics can no longer accept".[8] And Pick criticizes the very distinction between the terms "human" and "animal" as "conceptually and materially indecisive [...], a site of contestation, anxiety, and ritual [...], a zone in which the upkeep of human integrity, as it were, exacts a devastatingly violent price on animals".[9] So in the analysis below, I keep both the traditional (reformist, welfare-oriented) and the critical (radical hierarchical-rejectionist) positions in view.

Those moments in Spanish American literature that assert the importance of human moral duty toward nonhumans to ensure animal welfare and advance ideas to reform harmful treatment would qualify as literary expressions of TAS. While those moments that expose and condemn the intangible, indecisive, and sometimes violence-prone separation of "human" and "animal" into separate and absolute categories express ideas that fit more readily within the parameters of CAS. Advancement of the ideas from these two positions, and perhaps even some easing of the tensions between them, may be achieved through what I have elsewhere called fauna-critical analysis of the representation of animals in literary fiction.[10] In short, this approach aims to contribute to the resolution of debates within the wider fields of animal studies; it reexamines literary history to affirm the importance of a broadened canon; and it recognizes and highlights

---

7   DeGrazia, David: *Taking Animals Seriously: Mental Life and Moral Status*. Cambridge University Press: New York 1996, p. 76.
8   Cavalieri, Paola: *The Death of the Animal: A Dialogue*. Columbia University Press: New York 2009, p. 33.
9   Pick, Anat: *Creaturely Poetics: Animality and Vulnerability in Literature and Film*. Columbia University Press: New York 2011, p. 1.
10 DeVries, Scott: *Creature Discomfort: Fauna-criticism, Ethics, and the Representation of Animals in Spanish American Literature*. Brill: Leiden 2016, pp. 24–32.

what can be uniquely achieved for humans and nonhumans through literary formats rather than through nonfiction texts, such as philosophical treatises or political manifestos.

The Short Stories of Horacio Quiroga

As I suggest above, therefore, any discussion of sentient animals from the Spanish American literary tradition profitably begins with Quiroga. His tales are inspired by many of the author's own experiences as a small-tract farmer in the Misiones region of Argentina and are narrated from unique perspectives: impoverished laborers, solitary pioneers, foreigners, and especially animals, such as dogs, tigers, turtles, stingrays, snakes, etc. His fictional work anticipates several of the current concerns within TAS and CAS, but I have also chosen to foreground Quiroga here for the way in which his stories frame the selection of novels and other texts that I consider below: a nineteenth-century novel *El moro* (1897) by Colombian José Manuel Marroquín, short stories by the Argentine *modernista* Leopoldo Lugones, an example of Regional literature in Nicaraguan Hernán Robleto's *Una mujer en la selva* (1936), an *indigenista* text, *Los perros hambrientos* (1939) by Peruvian Ciro Alegría, and Chilean Luis Sepúlveda's more recent *Un viejo que leía novelas de amor* (1989).

In the stories by Quiroga, the use of the perspective of animals as a narrative focus serves to provide the natural world with a kind of agency that resists and protests harmful activities by humans. In Quiroga's case, his personal struggle with nature to carve an existence from a wholly hostile landscape like the Misiones region of Argentina and his familiarity with that landscape have been thought to impart a certain ecological flavor to the representation of the nonhumans that appear in his short stories. French observes that the narrativization of the animal perspective allows Quiroga "to transcend the abstractions and valuations [of] nature [...] and to conceptualize nature in such a way that humans are not excluded from or set against the environment but symbiotically located within it".[11] And Rivera-Barnes specifically addresses how the attribution of sentience to animals represents the way in which his works "bring awareness to the fragility of the natural world".[12]

---

11 French, Jennifer: *Nature, Neo-colonialism, and the Spanish American Regional Writers*. Dartmouth University Press: Hanover, NH 2005, p. 69.
12 Rivera-Barnes, Beatriz: "Yuyos Are Not Weeds: An Ecocritical Approach to Horacio Quiroga". *ISLE: Interdisciplinary Studies in Literature and Environment* 16(1), 2009, p. 48.

Such ideas are well-suited to an analysis of one of Quiroga's most well-known stories, "Anaconda" (1918), a text that expresses the ecological concern, put in the mouths of sentient snakes, that the appearance of humans in the area will bring contamination and death: "ever since the beginning of time, man and destruction are words that mean the same thing for all Animalkind".[13] And in "El regreso de Anaconda" (1926), one snake observes that "mankind was, is, and always will be the cruelest enemy of the jungle".[14] The destruction of non-human habitats is central to animal protectionist ethics, but Quiroga's stories, along with the other texts considered here, communicate this idea by having animals (rather than scholars or activists) denounce the devastation.

The change of narrative perspective from human to animal that is typical in Quiroga's work elicits empathy for the experience of animals. While the use of animals to narrate or focalize stories can deny their otherness as it does in fables or fairy tales, another way to understand what is going on in Quiroga is that his stories are more an empathetic identification with animals rather than appropriation. As Vint observes, such representations lead us to reconsider the entire logic of the human-animal boundary and the role of this species division in preserving a category of those whose subjectivity might be ignored. Rethinking such a boundary altogether can take us "beyond strategies such as 'humanizing' and 'anthropomorphizing'—and their dark flipside, 'animalizing'—and toward the more difficult work of thinking through the ethics of multispecies community".[15]

Quiroga's turn to animal perspectives sometimes functions simply as a literary device with which to frame human experience, culture, and society. Yet for nonhumans, the empathetic configuration of emotional manifestations as inherent to whatever form their consciousness may take implicitly advocates for the attribution of moral status to such creatures. Quiroga does not go there, but as we shall see in the examples from each of the traditional divisions of Spanish American into which I have organized the analysis below, both his stories and the representative text selected for each section suggest the parameters for such discussions.

---

13 Quiroga, Horacio: *Cuentos de la selva*. Muchnik: Barcelona 1999, p. 105. (Note: all quotations from Spanish-language primary texts are my own translations.)
14 Quiroga, Horacio: *Nuevos cuentos de la selva*. Solaris: Buenos Aires 1997, 2:132.
15 Vint, Sherryl: *Animal Alterity: Science Fiction and the Question of the Animal*. Liverpool University Press: Liverpool 2010, p. 80.

Nineteenth-Century Literature: *El Moro*

The narrative in Quiroga's "El potro salvaje" (1924) is configured as highly empathetic with animal experience and contains several moments that express access to the interiority of a young colt brought from the interior to race at the city horse track. The story concludes with this advice:

> Young colt: Hold strong to your career, although you can hardly make enough to eat. Because even if you pass away without achieving glory and even if you trade away your style for plentiful food, you will be saved because one day, you gave it all for nothing but a handful of hay.[16]

The horse loves the spectacle of the arena just for the thrill of racing and the narrator's expression of the importance of exertion for its own sake rather than for the promise of reward seems something like an inheritance from the didacticism of Spanish American nineteenth-century literature. That is, after the wave of independence movements in that century and region, the novels and poetry by authors from those new nations had incorporated into the plots and themes of their work idealized proposals for each new country's national identity. *El moro* by José Manuel Marroquín does something similar for the newly formed republic of Colombia, but its use of a horse for the narrator implies the relevance of such principles to nonhuman realms as well.

Marroquín's *El Moro* appeared at a time when horse stocks were widely-used in Colombia, but also about twenty years after the 1877 publication of the best-selling *Black Beauty* by England's Anna Sewell. Marroquín's criticism of poor animal treatment, elucidated "straight from the horse's mouth", echoes Sewell's novel and its didactic emphasis is typical of the Spanish American literature of that time. But from a fauna-critical perspective, there are several aspects within the novel that demonstrate that it does not just function as allegory for best human political practices, but that the text expresses several CAS ideas as well. In the novel, the horse / narrator complains about a cruel, sadistic owner characterized for his "love of making animals miserable"[17] and about the yanking of reins, the application of burning brands to claim ownership, the use of horses in war, and the imposition of intense labor for pack-work. Yet the peculiar focalization of the narrative from the perspective of a horse in *El Moro* with its unflinching emphasis on horses' capacity to experience pain, torture, and abuse also inevitably invokes several of the more philosophical concerns about

---

16 Quiroga, Horacio: *El desierto*. Losada: Buenos Aires 1997, p. 107.
17 Marroquín, José Manuel: *El moro*. Editorial Oveja Negra: Bogotá 1985, p. 63.

sentience and disruptions of the human / nonhuman hierarchical arrangements that might describe many of the tendencies within CAS as I have summarized them above. While scholars have never identified human moral responsibility for the ethical treatment of nonhuman fauna or criticism of the idea of the primacy of human subjectivity as ideological touchstones for nineteenth-century Spanish American literature, such elements, explicitly present in Marroquín's novel, suggest that this has been an oversight.

El modernismo: Leopoldo Lugones

Quiroga's "El mono ahorcado" (1907) is something of a sequel to "Historia de Estilicón", which I reference at the outset: the same deranged pseudo-scientist interested in primates narrates both stories. In the second text, the narrator probes the limits of animal sentience and ability to use language through tests that intend to measure the extent of pain that an animal can endure. In the story, the man subjects a monkey to a series of experiments aimed at getting him to speak, but the experiments fail miserably, as the man notes in his log:

> This morning I found him hanged. I tried the knot and found it secure and well-tied: I was convinced that the monkey's death was no accident. […]
> […] It might all have been because of my admittedly morbid and perverse curiosity or just the case of an animal-gone-wild after having been tortured for six months on end; though I cannot help but conclude that the intensity of a fully-human sense of anxiety unleashed in the monkey's brain was just too much. (Quiroga 1997, vol. 3, p. 38)

That the monkey seems to finally capture the abstract notion of anguish only to take its own life makes the tragic point that humans may cause so much distress to animals under their control that death is preferable.

In the short fiction of Leopoldo Lugones, there is a somewhat similar story, but from a slightly earlier Spanish American literary tradition. He was among a group of authors called the *modernistas*, writing from about 1880 to 1920, who can be characterized for their innovative use of language in poetry and for their poetic imagery in prose. The two most canonical figures—Nicaraguan Rubén Darío and Lugones—worked in both forms. In Darío's most well-known text, *Azul* (1888), the poem "Estival" features an idealized and tragic representation of "love" between two tigers. And while Darío's idealization of the tigers in "Estival" can make for a certain closing of emotional distance between animals and humans, fauna-critical analysis of animal representations in Lugones's stories reveals something quite different. His short-form fictional prose, especially in the collections *Las fuerzas extrañas* (1906) and *Los cuentos fatales* (1926), can be characterized for its explorations of the bizarre: the disconcerting capacities of scientific exploration, the nightmares of classical or biblical mythology, and the realms of the nonhuman. Lugones takes animals as the principal focus in at least two texts: "Los caballos de Abdera" and "Yzur"; below I consider the second of these.

In "Yzur", the concept of animals' capacity for language is explored as it was in "El mono ahorcado", but in this case, from the perspective of a narrator who wants to test a hypothesis about language use among apes. He keeps a monkey in his house and when the housekeeper says that she had heard the monkey produce several audible words, the narrator reports, "I realized right then and there, that he did not talk only because he did not want to",[18] but wonders, if monkeys are similar enough to human beings that their mental capacities are sufficient for speech, then why do they refuse to speak? In the story, the answer to this question comes from the apparent awareness among nonhuman primates that humans tend to act cruelly toward those considered less intellectually advanced:

> Despite a certain gloomy, barbaric, anthropoid misfortune for having delayed their evolutionary advancement, there was nevertheless an undeniable overthrow of the edenic, tree-dwelling, four-legged race, thinning their ranks, capturing their females so as to make slaves of the species right from the womb, and even prodding them, through the force of a sense of absolute impotency, to make a definitive break in their identification with humans as one and the same dignified species, preferring instead to remain mute and finding refuge in the darkness of animality. (p. 246)

The seemingly intentional nonhuman delay in "evolutionary advancement" or the recourse to "refuge in the darkness of animality" features something of the current CAS orientation toward a decentering of human sentience as categorically unique along with the moral obligation that this implies for humans, specifically the condemnations of speciesism. In the end, and even though ethical scrutiny of the character of nonhuman sentience was not the main concerns for *modernista* authors, stories like "Azur" disrupt reductive accounts of the human/animal distinction. By so doing, they constitute an implicit affirmation of undifferentiated ethical consideration, an animal-centric position that is, after fauna-critical analysis, implicitly present in Lugones's text.

---

18 Lugones, Leopoldo: "Yzur". In *Las fuerzas extrañas; Cuentos fatales*. Editorial Trillas: Mexico City 1992, p. 244.

Regional Literature: *Una mujer en la selva*

Quiroga's two simian stories, "Historia de Estilicón" and "El mono ahorcado", are fictionally unified in that each feature the same human keeper configured as narrator. They are also similar in that both conclude tragically, one with unspeakable violence and the other with animal suicide. While Quiroga was still alive and finishing the last of his texts in the 1930s, a by now long forgotten novel similarly explores the relationship between a human and ape that live together: *Una mujer en la selva* by Hernán Robleto. The novel, however, paints a far brighter picture for the possibilities of peaceful human/animal coexistence. With tones of the film *King Kong*, released in 1933 three years before the novel's publication, *Una mujer en la selva* similarly features a woman, Emilia Rivera, captured by an enormous monkey who lives in a forest near the *hacienda* where she lives. And like Fay Wray's Ann Darrow, Emilia eventually grows to love the animal whom she has named Jongo. However, the return to civilization only occurs after Jongo dies of old age, and Emilia's now wild appearance terrifies those in the first village she approaches. Emilia must return to the forest and only stops long enough to steal a pad of paper as she flees the village; its pages, filled with the record of her time with Jongo and left in the jungle cave where she had lived, function as a "found manuscript", the literary frame of the novel.

This text makes a case for the sentience of animals, as we have seen in *El moro* and the stories by Quiroga and Lugones, but here, the idea is less about the possibility of animals using language and more about the relationship between sentience and identity. By the end of the novel, Emilia assumes what Regan might call Jongo's psychological unity for the ways in which his behavior corresponded to her own. As Emilia begins to grow accustomed to the ape, she develops a true sense of companionship with him, and her journal repeatedly affirms his rational intelligence and even suggests that he has a soul. In one instance, Emilia considers the ape's sense of mental awareness: "The great primates concentrate their attention on deeper and more complex things. Jongo takes care of me and I know that I am important to him, even when I am not physically close. […] I make up a key concept in Jongo's mental horizon".[19] This moment is key in the development of Emilia's understanding of Jongo as an individual and of the possibilities for other animals' identity and sentience. DeGrazia develops a similar argument, emphasizing possible similarities between human and animal mind states:

---

19 Robleto, Hernán: *Una mujer en la selva*. Editorial Ercilla: Santiago de Chile 1936, p. 91–92.

Many animals have minds whose contents are not wholly dissimilar to the contents of human minds. Therefore, we need human phenomenology to study animal minds. Specifically, human phenomenology sets an agenda of what kinds of mental states to look for in animals and provides a start in understanding the qualitative features of animal mentation. (p. 78)

In Emilia's diary, the representation of Jongo's manner of thinking, her description of his capacity for long-term memory and concept formation, and the idea that their relationship forms part of the animal's identity are all similar to the way in which a human might process such things.

The preceding analysis traces the contours of a TAS perspective: inasmuch as animal sentience is similar to human capacities, philosophers like DeGrazia, Singer, Regan, and perhaps even Bentham, would consider that sufficient for moral duty. But from the perspective of CAS, the deconstruction of the categories like "human" and "animal", which is also an element of Robleto's text, is more central. For example, from the latter entries in her journal, Emilia's attention to the intelligence of Jongo begins to imply that simplistic categories such as "human" and "animal" no longer obtain. In one passage, for example, Emilia's reflections invite a reform of conventional thinking about animals along these very lines:

> What I am living, here on the margins of all human experience, only confirms the tyranny of conventional thinking. These are those mental rules that seek to dominate reason and do not allow it to spontaneously spread its wings.
> What I mean to say is that I begin to feel a kind of love for this animal I thought so horrible in the first weeks of my captivity. (p. 55)

This "tyranny of conventional thinking" informs the assumptions that Emilia has now come to question: the idea that reason must always bow to convention explains why categories such as "humanity" and "animality" have remained mutually exclusive. It is this separation that is so insistently challenged throughout Robleto's novel and, several decades after its publication in the 1930s, has now become a common target for criticism from the perspective of CAS.

Agamben, for example, emphasizes the precarious notion of such exclusions when he observes the "the division of life into [...] animal and human, therefore passes first of all as a mobile border within living man, and without this intimate caesura the very decision of what is human and what is not would probably not be possible".[20] And indeed, as this mobile border slips beyond that of Emilia to

---

20 Agamben, Giorgio: *The Open: Man and Animal*. Stanford University Press: Stanford 2004, p. 15.

include aspects of the monkey Jongo, the distinction is disrupted. Several times in the text, for example, Emilia questions her own identity. After Jongo dies, she makes a failed attempt to return to "civilization" and it fills her with dread. When she sees signs of other humans, she expresses ambivalence: "They are like cords that pull me in, but transport me to a kind of worrisome plane: worry that becomes torment if I follow the deductive conclusions suggested by my doubts. What if I'm not human?" (p. 140) An explicitly outright challenge to the very categories "human" and "animal" was something perhaps too radical for the novel's moment of publication: in the philosophical circles of that time, it is likely that Descartes's notion of animals as automatons still held sway. Nevertheless, by placing a challenge to these "conventions" in the fictional voice of a woman kidnapped by an ape and who later writes her story after rejecting the possibility of a return to human community, the novel creates sufficient distance so the challenge can at least be heard (or read).

Indigenista literature: *Los perros hambrientos*

In Quiroga's "La insolación" (1917), a man who has planted a cottonfield in Misiones is confronted by the spectral figure of death, which is only perceptible to his dogs. The story unfolds from their perspective as they attempt to drive it off, but the dogs know only too well that time is short for their "owner". As they face the prospect of losing the man who had taken good care of them, they begin to howl: "the four older dogs—so well-fed and so highly regarded by the owner that they were soon to lose—gathered together in the light of the moon, and with their snouts held aloft and their throats filled with sadness, howled out their cries of lament".[21] And as the appearance of death foretells, the man overexerts himself and dies in the unforgiving heat of that region; the dogs are distributed among the local peasants to suffer the hunger, want, and disease they had foreseen when they first considered that death had come for their master. A similar kind of animal suffering narrated from the perspective of the dogs themselves is featured in Ciro Alegría's *Los perros hambrientos*.

While there are several *indigenista* texts that feature animals, including one of the earliest, *Raza de bronce* (1919) by Bolivian Alcides Arguedas, perhaps the most relevant for consideration here is Alegría's novel for its representation of the plight of nonhuman, especially canine, life as it was experienced in the context of an indigenous community in the highlands of Peru. For

---

21 Quiroga, Horacio: *Cuentos de amor, de locura y de muerte*. Libresa: Quito 1989, p. 84.

literary scholars, the allegorical interpretation of the eponymous hungry dogs' anguish has long been the standard interpretation of the novel.[22] However, there are moments when the suffering of dogs is represented as more specific and unique than metaphorical; that is, the figurative plane does not work as allegory for the victimization of indigenous communities because what the dogs endure does not always function symbolically. In several instances, for example, the circumstances of the dogs' reactions to suffering are distinct from what humans would do, as in this case:

> The morning sun revealed the dogs lying on their side. They warmed themselves in its rays as they whined and drooled. Wanka had just given birth and tried to feed her litter of four pups, resigned to allowing them to nurse. The puppies, sickly and moving like small larva, seemed to be sucking her very blood.
> As day continued to dawn, the dogs began to wander about. Before so agile with the blood of the Alco breed in their veins, they could now hardly walk. They seemed nothing more than bags of bones with fur here and there. [...]
> [...] As night fell, a tragic chorus filled the highlands. The howling began and cut the silence like a knife. The dogs' lament seemed to mix in interminable sorrow. The wind would now carry it away, but their whines and barks and howls only emerged again and again from between a thousand hopeless jaws.[23]

If we look for allegory, the animal suffering represented here makes that comparison somewhat forced: motherhood and nursing as comparable to being sucked by leech-like "larvae" seem ill-suited as an adequate representation of the human experience of lactation. The dogs' howling for lack of nutrition is also unlike the human case: when individuals endure food scarcity, it is rare to waste energy to loudly complain about the fact. But howling behavior by dogs is a common response to stress: the pack must reunite and pool its resources to face adversity.

Los perros hambrientos features communal suffering, but of a type that transcends human experience. The attention to the unique nonhuman response to extreme adversity affirms the validity of canine perception as distinct from the parameters of human reactions in similar circumstances. On this point, Dawkins observes that:

> much of our behavior towards other people is [...] based on the unverifiable belief that they have subjective experiences at least somewhat like our own. It seems a reasonable belief to hold.

---

22 See Meléndez, Concha: Review of *Los perros hambrientos*, by Ciro Alegría, *Revista Iberoamericana* 3, 1941, pp. 226–228.
23 Alegría, Ciro: *Los perros hambrientos*. Cátedra: Madrid 1996, pp. 259–260.

Then we come to the boundary of our own species. No longer do we have words. No longer de we have the high degree of similarity of anatomy, physiology and behavior. But that is no reason to assume that they are any more locked inside their skins than are members of our own species.[24]

Human beings lack a similarity of consciousness that unfailingly lines up with nonhuman subjectivity. But the narrative representation of animal perception in *Los perros hambrientos* allows human readers a glimpse at what it is like to suffer hunger from the perspective of, for example, a mother dog who is tragically "resigned" to diverting nutrition from her own body to feed the pups or from that of a howling pack of hungry dogs that react so unlike human tendencies in the face of malnutrition. It is in these examples that the animal-centric achievement of Alegría's novel is manifest for the way in which it provides readers the opportunity to uniquely perceive certain qualities that characterize the peculiar aspects of nonhuman suffering.

The post-Boom: *Un viejo que leía novelas de amor*

In Quiroga's "El paso del Yabebirí" (1918), a settler near the Yabebirí river in Misiones prohibits the use of dynamite fishing near his property: "he didn't want to allow the senseless killing of millions of fish" (Quiroga 1999, p. 79). As is typical in a Quiroga story, the stingrays that live in the river are sufficiently sentient to perceive this kindness by the man and respond by defending him from cougars or jaguars that attempt an attack. *Un viejo que leía novelas de amor* by Luis Sepúlveda also features an antagonistic and fully sentient feline: in this case, a female ocelot that threatens a small jungle community after her mate and young have been hunted by an inept gringo. However, before I proceed to the analysis of this last sample of animal-oriented text from the most recent period of Spanish American literature, it will be helpful to provide some literary context.

While I have been using Quiroga's work for a frame of the texts selected for consideration here, I have also been tracing a roughly chronological progression from nineteenth-century fictions and *modernismo* to Regional and *indigenista* titles from the first half of the twentieth century. I now conclude with Sepúlveda's novel in this most recent literary period known rather clumsily as the post-"Boom", a designation that invites further clarification of the "Boom" itself. In an essay that seems in retrospect to be something of a manifesto for the Boom's

---

24 Stamp Dawkins, Marian: "The Scientific Basis for Assessing Suffering in Animals". In Singer, Peter (ed.): *In Defense of Animals*. Basil Blackwell: New York 1985, pp. 27–28.

literary tendencies, Carlos Fuentes defines it as markedly different from what had come before; that is, different from earlier novels that Fuentes denigrates as having been "swallowed by the mountain, by the plains, by the mines, or by the river".[25] It might not surprise, therefore, that "Boom" authors are not especially well-known for their representation of animals that might be found in such topographies. Furthermore, the "new" narrative emphases do not readily lend themselves to the representation of nonhuman beings; indeed, formal concerns about textuality, language, structure, and genre seem to preclude the explicitly ethical concerns that might be found in more activist literature such as nineteenth century foundational fictions or *indigenista* texts. And even those "Boom" novels that are set in the jungle—Peruvian Mario Vargas Llosa's *La casa verde* (1968) and *Pantaleón y las visitadoras* (1973), for example—are more concerned with formal literary innovations than with realist representations of fauna and flora.

As scholars began to think about literature from the mid-1970s to the present as a post-"Boom", some of the earlier literary tendencies from before the "Boom" started to reappear. There was something of a return to politically compromised novels by Luis Sepúlveda, for example, whose work emphasized ethical statements about ecology and animals. This is perhaps most apparent in *Un viejo que leía novelas de amor*, the text with which I will conclude the analysis here. The novel begins with the appearance of a dead *gringo* hunter near where the old man of the title—Antonio José Bolívar Proaño—has come to live out his days. Earlier, he had moved with his young wife to settle in the Amazon and eventually goes to live among the indigenous Shuar inhabitants of the region. From them, he learns how to survive and even thrive in the jungle. Many years later, Proaño returns to live among the settlers—the colony is now a town called El Idilio—and the dead *gringo* makes his appearance. It turns out he had bled out after being attacked by a female ocelot whose young he had killed and whose mate he had wounded. The rest of the novel deals with the events that unfold as a consequence of what has happened: the ocelot has tasted human blood and is a threat to the community. Proaño, the only one with enough jungle knowledge to be successful, reluctantly agrees to hunt the animal.

The animals in *Un viejo que leía novelas de amor* mainly appear in the context of hunting, which within the field of TAS is a complex and multi-faceted issue while for CAS, it is almost never countenanced as ethically permitted.

---

25 Fuentes, Carlos: *La nueva novela hispanoamericana*. Cuadernos Joaquín Mortiz: Mexico City 1969, p. 9.

From this latter perspective, Dunayer condemns the practice in no uncertain terms, but particularly when hunting is undertaken for leisure or entertainment purposes: "Categorizing hunting as a 'sport' or 'recreation' hides its violence and injustice inside a trivial context of play and completely discounts the nonhuman perspective".[26] Alternately, some TAS theorists may perceive some value in traditional hunting practices, such as when Fellenz holds that "hunting cultures afford epistemological space to the animal", and that "respect for the animal of which the hunter alone is capable comes from this intimate acquaintance, something of a fundamentally different order from human compassion for the prey".[27] Finally, Haraway recognizes the conflicts at play with regard to this issue when she affirms that "there is no way to eat and not to kill, no way to eat and not to become with other mortal beings to whom we are not accountable".[28] These questions concerning eating and hunting illustrate an ongoing tension in which the CAS position rarely recognizes the validity of hunting and never the use of animals for food while some more traditional, less radical positions allow that these may be more acceptable in certain, narrowly-defined circumstances. This tension also finds expression within the fictional space of *Un viejo que leía novelas de amor* with fictional episodes that affirm ethical principles more closely aligned to the CAS position than any other.

The bulk of Sepúlveda's novel deals with the search for the ocelot because it represents a serious threat to the community at El Idilio. Yet although recreational hunting of the ocelot species for sport is roundly criticized throughout the text, other forms of hunting are never categorically condemned in the novel; instead, *Un viejo que leía novelas de amor* explores several moments when exceptions may (and sometimes must) be made. As Proaño and the ocelot circle each other in a prelude to the novels denouement, the man reflects, in a monologue addressed to himself, about his previous experience and the principles that inform his hunting: "ocelots are not unfamiliar to you; it is just that you have never killed a cub, neither of ocelot nor any other species. Only adults as indicated by the laws of the Shuar".[29] Later, he observes that "the Shuar do not hunt ocelots. Their meat is inedible and the skin of just one is enough to

---

26 Dunayer, Joan: *Animal Equality: Language and Liberation*. Ryce Publishing: Derwood, MD 2001, p. 50.
27 Fellenz, Marc: *Moral Menagerie: Philosophy and Animal Rights*. University of Illinois Press: Champaign 2007, pp. 216–217.
28 Haraway Donna J.: *When Species Meet*. University of Minnesota Press: Minneapolis 2008, pp. 295, 299.
29 Sepúlveda, Luis: *Un viejo que leía novelas de amor*. Tusquets: Barcelona 1993, p. 122.

make hundreds of adornments that will last for generations" (p. 123). Finally, his thoughts turn to the relatively even odds that he faces in the stand-off with the great cat: "Why do you remember all of this now? Why are you so distracted by this animal? Maybe because you both know it will be an even match? After killing four people, she knows a lot about humans, as much as you do about ocelots. Or maybe you know less" (p. 123). Inherent in these reflections is the expression of a position that TAS scholars may invoke to justify hunting (or at least in which they refrain from condemning it): that the young of a species are not to be killed and that hunting is forbidden when no appreciable gain can be gotten from killing the animal. This orientation aims to reform hunting so as to make it more sustainable, yet because it does not advocate for absolute abolition, this position would be rejected from the CAS perspective.

But the third quotation from the previous paragraph indicates that for this contest between Proaño and the ocelot, a state of affairs exists that rarely or never does in the course of a recreational hunt: who will be the victor is not a foregone conclusion. Dunayer observes that

> in sport hunting, fair chase doesn't exist. Fairness requires comparable consequences for all "players." [...] A sport hunter who "loses" suffers only some disappointment, until his or her attention shifts to some harmless activity or another potential victim. But if the human wins, the nonhuman dies. (p. 49)

Yet in the novel, who chases and who is being chased is never clearly defined and the stakes for all involved are death: either the man or the animal will walk away. And Proaño very nearly loses his life; he only survives by hiding under an overturned canoe and must shoot through his own foot to kill the animal who has begun to dig through the earth to get at him. So even from a CAS perspective, this hunt may pass ethical muster, but only because it becomes a question of self-defense. Yet even so, most CAS scholars would likely still reject the premise of an ethically-justified hunt, even in a case that might be similar to the story from *Un viejo que leía novelas de amor*.

Still, the novel dramatizes the tension among reformist TAS positions, ambivalent views like Haraway's, and the more radical CAS stance. When Proaño finally kills the ocelot, he experiences profound sorrow rather than the thrill of victory: "The old man caressed her, ignoring the pain from his wounded foot, and he cried, ashamed, feeling unworthy, filthy, in no way the winner of that battle." (p. 136) The sadness and disgust he experiences is nothing like the adrenaline-fueled rush of the "sportsman" who employs technological aids to kill an animal; rather, the old man is reluctant, resigned to undertake the hunt only because the ocelot herself has begun to kill other sentient creatures. And Proaño's lament invokes many of the themes around which animal studies, both traditional and critical, organize advocacy: species extinction, the value of sentient nonhuman animals, and the disruption of the sense that humans are hierarchically superior to other species. But what seems most profound in the representation of Proaño's sorrow is that his deadly interaction with the ocelot only had to come to that because of the ecological ignorance of the long-dead *gringo*. The novel, therefore, only resolves a violent state of affairs unleashed by the killing of a highly sentient animal, by representing the ocelot's death as morally condemnable, an ethical position that fauna-critical analysis reveals as an inevitable conclusion from both the radical CAS and reformist TAS positions.

In the previous several paragraphs, I have attempted to provide examples of the way in which the ethical representation of animals in Spanish America's literary periods anticipate issues of concern within the broader field of animal studies, both critical and traditional. These include rights and sentience as well as the disruption of the validity of categories like "human" and "animal". I have assembled representative examples of Spanish American narrative fiction that address these issues and if not for the constraints of space, I could have included many more. But as we have seen in the texts by Quiroga, Marroquín, Lugones, Robleto, Alegría, and Sepúlveda, there is a rich if not particularly well-known subset of animal-centric, animal-ethical narrative in this literature. And it is a tradition that, as the importance of moral duty toward members of species other than our own becomes more of a pressing issue in our societies and cultures, deserves to be revisited by scholars, students, activists, and casual readers alike.

Bibliography

Agamben, Giorgio: *The Open: Man and Animal*. Stanford University Press: Stanford 2004.

Alegría, Ciro: *Los perros hambrientos*. Cátedra: Madrid 1996.

Bentham, Jeremy: *The Principles of Morals and Legislation*. Hafner: New York 1948.

Cavalieri, Paola: *The Death of the Animal: A Dialogue*. Columbia University Press: New York 2009.

DeGrazia, David: *Taking Animals Seriously: Mental Life and Moral Status*. Cambridge University Press: New York 1996.

DeVries, Scott: *Creature Discomfort: Fauna-criticism, Ethics, and the Representation of Animals in Spanish American Literature*. Brill: Leiden 2016.

Descartes, René: *Discourse on the Method*. In: Adler, Mortimer J. (ed.): *XXXI, Great Books of the Western World*. Hutchens, Robert Maynard (series ed.). Encyclopaedia Britannica: Chicago 1952.

Dunayer, Joan: *Animal Equality: Language and Liberation*. Ryce Publishing: Derwood, MD 2001.

Fellenz, Marc: *Moral Menagerie: Philosophy and Animal Rights*. University of Illinois Press: Champaign 2007.

French, Jennifer: *Nature, Neo-colonialism, and the Spanish American Regional Writers*. Dartmouth University Press: Hanover, NH 2005.

Fuentes, Carlos: *La nueva novela hispanoamericana*. Cuadernos Joaquín Mortiz: Mexico City 1969.

Haraway, Donna J.: *When Species Meet*. University of Minnesota Press: Minneapolis 2008.

Huggan, Graham / Tiffin, Helen: *Postcolonial Ecocriticism: Literature, Animals, Environment*. Routledge: London 2010.

Lugones, Leopoldo: *Las fuerzas extrañas; Cuentos fatales*. Editorial Trillas: Mexico City 1992.

Marroquín, José Manuel: *El moro*. Editorial Oveja Negra: Bogotá 1985.

Meléndez, Concha: Review of *Los perros hambrientos*, by Ciro Alegría. *Revista Iberoamericana* 3, 1941, pp. 226–28.

Pick, Anat: *Creaturely Poetics: Animality and Vulnerability in Literature and Film*. Columbia University Press: New York 2011.

Quiroga, Horacio: *Cuentos de la selva*. Muchnik: Barcelona 1999.

Quiroga, Horacio: *El desierto*. Losada: Buenos Aires 1997.

Quiroga, Horacio: *Nuevos cuentos de la selva*. 3 volumes. Solaris: Buenos Aires 1997.

Quiroga, Horacio: *Cuentos de amor, de locura y de muerte*. Libresa: Quito 1989.

Regan, Tom: *Animal Rights, Human Wrongs: An Introduction to Moral Philosophy*. Rowman and Littlefield: Lanham, MD 2003.

Rivera-Barnes, Beatriz: "Yuyos Are Not Weeds: An Ecocritical Approach to Horacio Quiroga". *ISLE: Interdisciplinary Studies in Literature and Environment* 16(1), 2009, p. 35–52.

Robleto, Hernán: *Una mujer en la selva*. Editorial Ercilla: Santiago de Chile 1936.

Sepúlveda, Luis: *Un viejo que leía novelas de amor*. Tusquets: Barcelona 1993.

Singer, Peter: "Ethics, Animals and Nature". In: Li, Hon-Lam / Young, Anthony (eds.): *New Essays in Applied Ethics: Animal Rights, Personhood and the Ethics of Killing*. Palgrave: New York 2007, pp. 29–41.

Singer, Peter: *Animal Liberation*. Avon Books: New York 1977.

Stamp Dawkins, Marian: "The Scientific Basis for Assessing Suffering in Animals". In Singer, Peter (ed.): *In Defense of Animals*. Basil Blackwell: New York 1985, pp. 27–40.

Vint, Sherryl: *Animal Alterity: Science Fiction and the Question of the Animal*. Liverpool University Press: Liverpool 2010.

Arturo Arias

Indigenous Knowledges and Ecological Thought: Jak'alteko Maya Victor Montejo's Fables

Abstract: *The Bird Who Cleans the World and Other Mayan Fables* is, in the words of the Jak'alteko Maya writer Victor Montejo, "a testimony to the values of respect, unity and understanding that existed between the people and their natural and supernatural environment" (pp. 15–16). It is also a validation of "animal subjects" and their interaction with human subjects in an integrated eco-space where everything is a living subject. That is, people and animals, plants and springs, clouds and caves, light and wind, hills and valleys, stones and rivers, pots and griddles, crosses and roads, to rephrase Carlos Lenkersdorf, an ethnographer who worked and lived with the Tojolab'al Maya of Chiapas, Mexico. Montejo's animal characters are also other-than-human persons, social beings participating in world-making relations among themselves and with humans, one that closes the divide between nature and culture. Montejo's work evidences how traditional ecological knowledge is both highly localized and social. The focus is located on the web of relationships between humans, animal, plants, natural forces, spirits and landforms of his specific locality. These fables explain the web of relationships specific to a place and to a group of people, describing their ecosystem as a sensuous connection of knowing between humans and nonhumans, where the natural world is not simply passive. Elements are active agents that also embody ways of knowledge and participate in a negotiated order, which is also an affective and transformational mode of bodily knowing. Meaning thus results from a cultivated bodily attention to the environment.

Keywords: Víctor Montejo, Mayan Fables, Indigenous Knowledge, Ecocriticism, Animal Ethics

This chapter analyzes Montejo's *The Bird Who Cleans the World and Other Mayan Fables* to problematize ongoing debates regarding ecological spaces, approaches to animal and plant studies that have derived from it, and Indigenous ontological differences (Viveiros de Castro's phrase) that challenge and rupture traditional Western understandings of the subject and his/her world and, in so doing, consider how animal and plant studies have emerged in the Global North. The conflict between these movements and traditions relate to ontological politics. Modern globalized technologies threaten ecological spaces and offer vision where "nature" becomes an object to be controlled for profit. Responses to this threat have been global. Nonetheless, most articulated from Western perspectives only pay lip-service to the global/local divide, even though it has been nearly two decades since Argentinian scholar Walter Mignolo stated in *Local Histories / Global Designs* that we need to consider "the *location* of the agency and locus of enunciation from where imaginary constructions [...] are produced", and what he names "the languaging (rather than language) differences between the two Americas".[1]

Argentinian critic Horacio Legrás argues along these lines for a "communitarian ethos" that challenges "the enlightenment's narrative that locates the origin of society in a differentiation from the natural state".[2] I posit that this communitarian ethos and location of agency are manifested in the languaging that configures contemporary Indigenous imaginary constructions. Yet, ecological perspectives and animal and plant theories crafted in the Global North always ignore these perspectives. This chapter is an invitation to examine the limits of the modern nature / culture divide and rethink the historical results of colonization, by considering what animal theory may be from a Maya perspective.

New Disciplines and Their Debates

Ecological perspectives and new configurations of animals and plants as active beings and agents with communicative organisms have appeared in the Global North since at least the last quarter of the twentieth century. Their sources vary, from Deleuze and Guattari's rhizomatic positionalities, which launched many of these issues, to Jacques Derrida's ethics of hospitality, or Isabelle Stengers's and Bruno Latour's efforts to democratize science, even if the two of them do not represent the same theoretical stance. In Deleuze and Guattari's case, this argument appears as an "anti-humanistic approach for which the animal behavior becomes exemplary in its capacity to express the power of an impersonal life".[3] It is a complex debate located outside the purview of the present essay. In this chapter, I merely outline a rough sketch of some of these tendencies in the Global North.

---

1 Mignolo, Walter D.: *Local Histories/Global Designs: Coloniality, Subaltern Knowledges, and Border Thinking*. Princeton UP: Princeton 2000, p. 193.
2 Legrás, Horacio. "The Rule of Impurity: Decolonial Theory and the Question of Literature". In: Juan G. Ramos and Tara Daly (eds.): *Decolonial Approaches to Latin American Literatures and Cultures*. Palgrave Macmillan: New York 2016, p. 23.
3 Beaulieu, Alain: "The Status of Animality in Deleuze's Thought". *Journal for Critical Animal Studies* 9 (1), 2011, p. 72.

In "The Truth of Others: A Cosmopolitan Approach", German sociologist Ulrich Beck argues that the discourse of postcolonialism had effectively disrupted Eurocentric political and cultural forgetfulness.[4] He evokes the 1550 debate between Bartolomé de Las Casas and the Spanish royal historian Juan Ginés de Sepúlveda before the Council of the Indies in Valladolid, on the putative humanity or inhumanity of Indigenous peoples, to "help clarify what we ourselves are arguing at the turn of the twenty-first century" (p. 432). This debate launched the concept of race in Western consciousness, according to David Theo Goldberg in *The Racial State*.[5] In an understated manner, Beck cautions about the risk of cosmopolitics reintroducing undesired and mostly unintended Eurocentric tenets into what he labeled a "second modernity" (p. 430).

French sociologist Bruno Latour responds to Beck on the same issue in his own article, "Whose Cosmos, Which Cosmopolitics?".[6] Latour claims that "the limitation of Beck's approach is that his 'cosmopolitics' entails no cosmos and hence no politics either" (p. 450). He goes on to claim that society has never been limited to humans (p. 451). Berating Beck for not being able to go beyond the Valladolid debate to state his point, Latour responds to him via Brazilian Anthropologist Eduardo Viveiros de Castro. Latour states that what was truly at issue in Valladolid in 1550 was not whether Indigenous peoples had souls, but whether Conquistadors had bodies. He then proceeds to paraphrase Viveiros de Castro extensively:

> The theory under which Amerindians were operating was that [...] Entities all have souls and their souls are all the same. What makes them differ is that their bodies differ, and it is bodies that give souls their contradictory perspectives [...] Entities all have the same culture but do not acknowledge, do not perceive, do not live in, the same nature. For the controversialists at Valladolid, the opposite was the case but they remained blissfully unaware that there was an opposite side. (pp. 451–452)[7]

Latour informs readers that Indigenous peoples also doubted if the Conquistadors had spirituality, though they knew they had bodies and conducted experiments to find proof.[8] He goes on to state that Viveiros de Castro has shown that there are

---

4   Beck, Ulrich: Patrick Camiller (trans.). "The Truth Of Others: A Cosmopolitan Approach". *Common Knowledge* 10 (3), 2004, p. 449.
5   Goldberg, David Theo: *The Racial State*. Blackwell: Oxford 2002.
6   Latour, Bruno: "Whose Cosmos, Which Cosmopolitics?". *Common Knowledge* 10 (3), 2004, pp. 450–462.
7   Viveiros de Castro, Eduardo: "Cosmological deixis and Amerindian perspectivism". *Journal of the Royal Anthropological Institute* 4 (3), 1998, pp. 469–488.
8   "Conquistador prisoners were taken as guinea pigs and immersed in water to see, first, if they drowned and, second, if their flesh would eventually rot. This experiment was

many ways to be "other" (p. 453), before arguing on behalf of Isabelle Stengers's mode of cosmopolitics (p. 454):

> The presence of *cosmos* in *cosmopolitics* resists the tendency of politics to mean the give-and-take in an exclusive human club. The presence of *politics* in *cosmopolitics* resists the tendency of *cosmos* to mean a finite list of entities that must be taken into account. *Cosmos* protects against the premature closure of *politics*, and *politics* against the premature closure of *cosmos*. (p. 454)

In a nutshell, Latour's summary of Stengers's work has become the launching pad for new theoretical approaches about nonhuman subjects. Cosmos signifies all the nonhuman entities "making humans act" (p. 454). The last phrase remains, nonetheless, problematical, given the implication that nonhuman agency exists solely to make human subjects act. We will return to this point later.

In "Cosmopolitics and the Subaltern", anthropologist Matthew C. Watson points out that Latour's project fails "to apprehend the experiences of subjects marginalized by scientific and political structures of representation".[9] He reads Latour "against Dipesh Chakrabarty's concept of 'subaltern pasts', which are unassimilable to academic historical narratives" (p. 55). Watson concludes that Latour "does not provide any tools to understand representational elisions" (p. 73). In a subsequent article "Derrida, Stengers, Latour, and Subalternist Cosmopolitics", Watson reconciles Derrida's trace of the limits of the knowable with Stengers and Latour.[10] Concerned by cultural anthropologists' engagement with "otherwise-voiceless collectives and subjects on a daily basis" (p. 76), Watson develops a "vision for a subalternist cosmopolitics, an ontological politics that attends to its limits of representation and the forms of violence that its boundary-practices enact" (p. 76). Explaining Western scholars' tendencies to ignore subalternized and racialized subjects' experiences, narratives, and forms of being, Watson reformulates his argument that "marginal or abject actors—the subaltern—" are "unrepresented within and even unrepresentable by contemporary institutions of politics and science" (p. 93). Needing "spokespersons" in both academia or society in his understanding, academics should attach and

---

as crucial for the Amerindians as the Valladolid dispute was for the Iberians. If the conquerors drowned and rotted, then the question was settled; they had bodies. But if they did not drown and rot, then the conquerors had to be purely spiritual entities, perhaps similar to shamans" (p. 452).

9 Watson, Matthew C.: "Cosmopolitics and the Subaltern: Problematizing Latour's Idea of the Commons". *Theory, Culture and Society* 28 (3) (2011): pp. 55–79.

10 Watson, Matthew C.: "Derrida, Stengers, Latour, and Subalternist Cosmopolitics". *Theory, Culture and Society* 31(1), 2014, pp. 75–98.

commit to their cause; however, the meaning of the latter resolution is left unexplained. In this logic, subalternized and racialized subjects still remain outside of Derrida's limits of the knowable. Care for "them" is exercised by Western academics familiar with their daily experiences and ontological understanding of the world. Watson rightfully advocates for their rights to asylum as they are, in many instances, persecuted subjects. The problem is that, even if offered radical hospitality in Derridean terms, these subjects remain as nameless foreigners to Western subjects in Watson's argumentation. Their metaphysical and ontological positionalities continue as equally unknown, except for the benevolent Western academic mediator who claims to "know" them. In this logic, we have not moved much beyond the late 1980s explanation by Spivak of why the subalterns cannot speak within Eurocentric parameters, nor away from Eurocentric liberal paternalism.[11]

Concrete subaltern subjects anchored in a specific location are missing in Watson's argumentation, together with their unique discursivities crafting singular social imaginaries. Animals, nonhuman subjects, have also been elided. Even the substance of Latour referential ties to Viveiros de Castro in his 2004 article disappears from Watson's eyes. Never do we see an argument transcending John Grim's affirmation that animals are typically understood as objects of utilitarian use in the West, that is, as entities "available to humans whether as food, entertainment, scientific research, or as pets", even if under some of these conditions they make humans exercise agency in Latour's terms.[12] For this reason, Grim affirms that "Indigenous cultures present a more seamless weave between social, economic, ecological and cosmological realms" (p. 378).

Perhaps the crux of the continuous absence—the ongoing condemnation of subalternized peoples and their ontologies to the realm of Derrida's limits of the knowable—remains the difficulty Eurocentric scholars manifest in explaining— perhaps justifying—indigenous ontologies in which nonhuman subjects interact with humans on the same plane. After all, despite Latour's defense of Stengers's approaches, nonhuman entities are still, for him, only present to "make humans act", as previously mentioned. The politics in cosmopolitics is still a chastity belt

---

11 Spivak, Gayatri Chakravorty: "Can the Subaltern Speak?". In: C. Nelson and L. Grossberg (eds.): *Marxism and the Interpretation of Culture*. MacMillan: Basingstoke: Macmillan 1988, pp. 271–313.
12 Grim, John: "Knowing and Being Known by Animals: Indigenous Perspectives on Personhood". In: Paul Waldau and Kimberly Patton (eds.): *A Communication of Subjects: Animals in Religion, Science, and Ethics*. Columbia UP: New York 2006, pp. 373–390.

to legislate human societies within Eurocentric parameters. Argentinian critic Horacio Legrás has already labeled this omission a "referential ingratitude", one that reflects Western scholars' reluctance to abandon Westernness. Despite the theoretical rhetoric, Legrás still sees a refusal to embrace the ontological positionalities of those cultures the West colonized (p. 20) and perhaps also an unwillingness to relinquish the position of power that represents the scholars' role as mediators between subalternized, racialized communities from the Global South and Eurocentric societies in the Global North. After all, if Western audiences were to listen to Indigenous scholars or writers, what would then become the role of Western mediators?

Working from Within Indigenous and Native American Positionalities

At the risk of generalizing, let me state that Indigenous ontologies do address modes of relation between humans and nonhumans, and they often do so by narrating or performing stories. It may also be done by rituals and ceremonies that gesture the mindsets, values, and experiences that organize communal meanings to gain knowledge and understanding of a concrete, specific place: their own habitat. They usually frame ontological meanings in creation stories in association with the explicit place where they live, have lived for millennia, and situate their cosmopolitical boundaries. Concrete uniqueness versus abstract categorization and taxonomy often establish an abyss between them and theorizations configured by Eurocentric philosophy and knowledge. Nevertheless, academics from the Global South or those of Native American or Indigenous origin have begun to work from within these perspectives.

In *Earth Beings: Ecologies of Practice Across Andean Worlds*, Peruvian anthropologist Marisol de la Cadena explains the continuity between human and nonhuman subjects in the interaction that led to confrontation in 2006 when Indigenous organizations from the Cusco region in Peru mobilized to prevent a mining operation in Ausangate, one of the most majestic peaks in the Andean chain.[13] There was specificity to this protest. The protest was neither a strictly anti-capitalist one, nor an ecological defense of the mountain in the classical sense. Rather, it was a cosmopolitical one. Ausangate is for Kechwas an *apu*, a mountain that is alive and therefore possesses *camay*, a spirit that is a life force, a supernatural vitalization of all material things. Indigenous communities feared

---

13 Cadena, Marisol de la: *Earth Beings: Ecologies of Practice Across Andean Worlds*. Duke University Press: Durham 2015.

that Ausangate could in consequence destroy them if the mining operation advanced. This was because it would interrupt millennial practices and ceremonies of respect, recognition, and affect that marked not only their relationship with the *apu*, but all the ones that existed between Indigenous communities and Andean Earth beings since the beginning of time.[14]

We have analogous examples of relationships between humans and nonhumans from the Tupi and Araweté of the Brazilian Amazon from anthropologist Eduardo Viveiros de Castro, cited positively by Latour. For over twenty years, Viveiros de Castro has been addressing what he has labeled the *ontological turn* in anthropology. In his ethnographic work, Viveiros de Castro claims that differences between viewpoints should be understood in ontological rather than epistemological terms. In his understanding, distinct species do not see the same world in diverse ways. Rather, they "see in the same way […] different things [or 'worlds']" (p. 478). All beings have a consciousness and a culture, and all see themselves as "human" and all others as "nonhuman". In this sense, difference becomes what things could be ontologically, rather than how they might be represented or constructed. Ultimately, this logic implies as well that nature is not what has been traditionally understood in Eurocentric thinking, rather, nature is a cultural construct.

In "Animal Bodies, Colonial Subjects: (Re)Locating Animality in Decolonial Thought", Native American Cree scholar Billy-Ray Belcourt claims that

> Critical Animal Studies (CAS) and mainstream animal activisms have failed to center an analysis of settler colonialism and therefore operate within "the givenness of the white-supremacist, settler state" […]. This theoretical absence is thus a form of colonial violence wherein indigeneity is invisibilized, wherein the Indigenous body is remade into a site of modern impossibility to make possible the re-shaping of animal subjectivities as settler-colonial imaginaries.[15]

---

14 Ausangate is one of the twelve sacred apus of Cusco. The others are Salkantay, Mama Simona, Pillku Urqu, Manuel Pinta, Wanakawri, Pachatusan, Pikchu, Saksaywaman, Viraqochan, Pukin, and Sinqa. Frank Salomon states that "camay connotes the energizing of extant matter" and "it is a continuous act that works upon a being as long as it lasts" (p. 16). Thomas Cummins adds, "the Andean cosmological concept of camay […] can be considered the supernatural vitalization of all material things" (p. 182). According to Tamara Bray, it is an essence or a power (p. 358). Jeffrey Quilter understands it as a distribution of life force across reality that does not make a sharp distinction between animate and inanimate (p. 48).
15 Belcourt, Billy-Ray: "Animal Bodies, Colonial Subjects: (Re)Locating Animality in Decolonial Thought". *Societies* 5 (1), 2015, p. 2.

Belcourt concludes that animal ontologies should be relocated within decolonial thought. This was because of his claim that settler colonialism "re-makes animal bodies into colonial subjects to normalize settler modes of political life that further displace and disappear Indigenous bodies and epistemologies" (p. 9). As such, decolonization is only possible by reconfiguring animal ethics within Indigenous ontologies. Subalternized and racialized subjects from the Global South could not agree more with Belcourt's argumentation.

When incorporating de la Cadena's or Viveiros de Castro's cosmopolitical and ontological positionalities, Belcourt's inclusion of animal subjects within decolonial perspectives, or critiques of literary texts written by Indigenous subjects, the limits of Eurocentric animal theorization become apparent. If literary criticism consists of determining those problems posed through representation in texts by a given culture in a specific moment of their history, the literary critic has the epistemological imperative to render the writer's cosmopolitical and ontological positionalities in as fully an integral way as possible. This should be so even when the critic ignores the original language and is often unfamiliar with the ontological knowledges represented to depict social relations, cognition, kinship, religion, politics, etc. in given texts. The best policy for critics is to allow Indigenous peoples to define themselves and their worlds through their literatures' signifiers. I will offer a brief example of this process with my analysis of Victor Montejo's animal fables.

Victor Montejo and Indigenous Knowledges

A Jakaltek Maya born in the town of Jacaltenango (named Xajla' in his native language) in 1951, Victor Montejo is a native speaker of Maya Popb'al Ti'. He first studied to become a primary-school teacher at the Instituto Indígena de Varones Santiago in Antigua, Guatemala, from 1970 to 1972. Later, he miraculously survived a Guatemalan army massacre in the early 1980s, as he narrates in his second published book, *Testimony: Death of a Guatemalan Village* (1987). Montejo fled to Chiapas, Mexico. With the help of Guatemalan American writer Víctor Perera and US actress Jane Alexander, he moved to the US. He completed his BA at Bucknell University, an MA in anthropology at the State University of New York–Albany in 1988, and a PhD at the University of Connecticut in 1993. Besides a scholar, Montejo also became a writer, publishing novels, short stories, poetry, testimonios, and academic books in three languages—English, Spanish, and Popb'al Ti'. He is one of the most prolific writers in the Latin American Indigenous orbit.

In *The Bird Who Cleans the World and Other Mayan Fables*, Montejo tells his readers in the author's preface that his mother scolded him once by stating that "(T)he young people of today do not believe in their Mayan heritage. That is why they cannot understand and value them".[16] This reference alludes to a fast-paced Westernization process that began in the Guatemalan highlands in the early 1970s. Montejo's intellectual, creative, and affective process came from the desire to keep Maya "heritage alive" (p. 15). His book of fables is but one example of this process, though it is the one in which animal stories appear in greatest number. His fables emerge from an oral tradition dating back thousands of years that establishes moral values in much the same way that descriptive or comparative ethics operate, without prescribing a human comportment. We can understand the moral reasoning that lies behind the Jakalteko cosmovision, structuring the community's ethics. As in other Indigenous cases—without generalizing or normatizing, both dangerous paths—Jakalteko knowledge forms part of a unique cosmological order, circumscribed by its precise geography and ecospace. It is First Nations' scholars Mi'kmaw Marie Battiste and Chickasaw James (Sa'ke'j) Youngblood Henderson who remind us that "Indigenous knowledges is not a uniform concept among all Indigenous peoples; it is a diverse knowledge that is spread throughout different peoples in many layers".[17] They add that taxonomies are not part of Indigenous thought. Finally, they mention something that pertains to Montejo's writing, even if they had never heard of him. Indigenous knowledges are not codified in abstract, scholarly protocols, but are manifested by humor, observation, experience, social interaction, and "listening to the conversations and interrogations of the natural and spiritual worlds" (p. 36).

One of the best examples appears in the author's preface itself. Montejo tells his readers that when he was a child his mother would recount the story of a wild dove pleading for cotton cloth to protect her leg from the cold. When they asked her what the problem was, the dove would respond with the following:

> Mis, mis, k'uxumtoq tx'ow;
> tx'ow, tx'ow, holom b'itz'ab';
> b'itz'ab' mach xhchanik'oq kaq'e;
> kaq'e, kaq'e, ch'iniq'oq asun;
> asun, asun, ch'ok yinh sat tz'ayik;
> tz'ayik, tz'ayik, xhmak'nitanhoq chew;
> chew, chew, xhq'ahnitoq woqan. (13)

---

16 Montejo, Victor: Wallace Kaufman (trans.): *The Bird Who Cleans the World and other Mayan Fables*. Curbstone: Willimantic 1991, p. 11.
17 Battiste, Marie and James (Sa'ke'j) Youngblood Henderson: *Protecting Indigenous Knowledge and Heritage: A Global Challenge*. Purich Publishing: Saskatoon, 2000, p. 35.

[Cat, cat, that eats and eats mice;
mice, mice, that gnaw, chewing holes in walls;
walls, walls, that stop the wind;
wind, wind, that carries the clouds;
clouds, clouds, that shadow the sun;
sun, sun, that kills the cold;
and cold, cold, that hurts my leg]. (pp. 13–14)

We can see here that, to explain why her leg is cold, the dove articulates a holistic understanding of her environment. She is cold within a relational process representing an Indigenous epistemic regime that does not distribute differences along the nature-culture divide, creating a complex entity where everyone and everything—cats, mice, walls, wind, clouds, sun, and dove—all function as full-fledged subjects belonging to a system in which everything is interconnected. Consequently, their actions affect all stakeholders, who are thus bound as a community by how their singular actions have an impact on all other members of the same community. They all have both an interest and concern in articulating how this organization apprehends multiplicities. This simple ethical model articulated by Montejo's mother allows for multiple, nonhierarchical entry and exit points (the dove could use other examples in complaining about the cold in her leg or begin in any other point of the many possible causal relations) while establishing transspecies connections. Repetitions of words, couplets, also illustrate scholar Nathan C. Henne's assertion that Maya writing depends heavily on couplets because parallel words evoke preHispanic glyphic writing in which sound phonemes and pictographs were conjoined. Thus, "only in the *juxtaposition of the two* [...] is the 'whole' description located".[18]

Montejo goes on to explain in his author's preface that in the past, birds were the "living colors of the world". The buzzard was especially appreciated because he always cleaned the villages' surroundings (p. 16), however, when the Guatemalan army invaded his region in 1982, besides massacring countless villagers, they used buzzards for target practice. This abject gesture thus broke one more connection in the Mayas' holistic understanding of how the environment, the world itself, is shared by all subjectivities, regardless of the nature of the species. Montejo adds, "Respect for nature has diminished to the point that modern people destroy their environment systematically out of thoughtlessness

---

18 Henne, Nathan C.: "A Cartography of the Uncertain: The Maya Textual Exile". In: Bishop, Karen Elizabeth (ed.): *Cartographies of Exile: A New Spatial Literacy*. Routledge: New York 2016, p.38.

or selfishness. People can destroy themselves by not recognizing the value of all living creatures on earth with whom they should coexist" (p. 16). He then affirms that, while Mayas "continue to be active producers of knowledge", their elders have been faulted for perpetrating "absurd ideas that are symbols of their 'backwardness'" (p. 18). Ultimately, Montejo claims that his work is, rather than an expression of the past, "a symbol of resistance" (p. 18).

The aspects highlighted in the previous paragraph are underscored by the fact that, as Allan Burns points out in the introduction to the volume, "the animal stories in Maya culture [...] are more substantial than the child-like animals of [...] other traditions. These Maya animals reveal profound character and personality traits, severe actions, and sometimes unpleasant ideas" (p. 22). He adds that people like Montejo's mother were "the shapers of Maya history and cultural theory" (p. 24), and that most stories in the volume are narrated in the dialogic form of conversations: "Characters do not so much engage in action as talk to each other. Their schemes, successes, and failures emerge from the talk they have with one another" (p. 25), with very little emphasis on action or setting. Burns's remarks emphasize the primacy of spoken interaction between subjects—whether human, animal, or divine—, a distinctive trait of many Indigenous cultures. This implies that intersubjectivity takes the shape of what Hirschkop labels "ubiquitous dialogism" in reference to Bakhtin, one in which the dialogical nature of utterances reveals the ethical patterns operating within a given conception of the world.[19] The dialogical nature of discursivity and the utterance generates contexts of speech that differentiate the Indigenous world from Western-centered configurations, as actual intersubjectivity is given shape and articulated within Montejo's representations. After all, in Hirschkop's understanding, "The shape of the utterance depends on the social sphere in which it is located within its characteristic worldview, topos, style, and vocabulary" (pp. 209–210), that is, within a specific ecospace that generates its singular identitary traits.

Montejo's collection contains thirty-two fables. The title story begins with a flood that "covered the whole world" (p. 29). As the water recedes, Usmiq, the buzzard, is sent to find out just how much the water had ebbed, and in his search, he discovers many dead and rotting animals. Starved, he begins to eat them. When he flies back, he stinks of the dead creatures he had eaten on the reappearing dry land. His job then becomes to "clean the world of stench and

---

19 Hirschkop, Ken: Mikhail Bakhtin. *An Aesthetic for Democracy*. Oxford University Press: Oxford 2000, p. 67.

rottenness" (p. 30), carrying off whatever might contaminate the land. The story thus evidences the buzzard's key role in the well-being of the world. And transforms an animal regarded as lowly in the Western world, a trope of lack of cleanliness in modern urban environments, into an ethical agent. In being chosen as the subject in charge of cleaning the earth's environment, the buzzard extends through this designation the traditional boundaries of ethics to include the nonhuman world. Undoubtedly embedded in subsistence practices that culturally validate personhood, the previous example also illustrates the mutuality of knowing between humans and animals. In the opinion of John Grim, a history of religions scholar, this knowing exchange is both affective and transformational (p. 373).

"The Little Boy Who Talked With Birds" addresses human / animal subject communication with a simple story. A child went every day to work with his unkind father in their cornfield (p. 67). At lunchtime, a bird would come "and perch […] in the tree above them" (p. 68). The bird would sing and the boy would sometimes laugh, smile, or remain pensive. One day his father finally asks him what the bird is saying. When the boy claims it is nothing, the father insists, and after many threats, the boy confesses that the bird had told him that one day his father would salute him (p. 68). The father becomes very upset, interpreting the boy's words as a warning that, as he grew up, he would disrespect him. He starts mistreating the boy more and more, and finally kicks him out of the house (p. 69). The boy begins wandering around like a lost child:

> When he had traveled for a long time, the boy came to the domain of a great chief where by chance he heard the following proclamation:
> *He who can interpret the squawks of the crows who come every afternoon fluttering about the chief's window, can marry the chief's daughter and inherit the kingdom.* (p. 69)

The boy comes before the chief and is given a chance to interpret what the birds were saying. The boy smiles and listens as he sits by the window where two crows come every afternoon and squawk. Then, he informs the chief that they are a male crow and a female crow fighting. The male claims the female had abandoned her eggs and he had to keep them warm until they hatched. She argues that the male had not brought any food to the nest and now wants to claim the two baby crows. The boy then recommends that the chief award the baby girl crow to the male, and the female should "take the little boy crow" (p. 70). The chief accepts this recommendation as do the crows when the boy explains the arrangement to them. The chief fulfills his promise and marries his daughter to the boy. When he inherits the kingdom, all the people from neighboring villages come to salute the new ruling couple. An old couple is among them, and the young chief rushes forward to greet the old man and says, "Don't bow before me and don't salute me, because I am your son" (p. 70). The old man sobs and apologizes:

> "Oh, my son! Forgive me for what I have done to you," the old man sobbed. The boy embraced his parents and announced, "Don't worry, father. I'm not angry. From today on, you and mother will live near me so our family can be reborn in peace and happiness." (p. 71)

The story ends on this note.

Rather than a more Western ecological approach, Maya cosmology explains this story. The child / bird dialogue relates to the concept of the *nagual*, or *nawal*.[20] For Mayas, a nagual is an animal correspondent for every subject and belongs to the *winaqil*, the soul / beingness, an animal double that every human subject has, determined by the calendar day on which they are born. In "Untranslation: The *Popol Wuj* and Comparative Methodology" Nathan Henne argues that it is "necessary to understand the real physical nature of *naguales*, which is often erroneously translated as 'animal spirits'. This translation is deceiving because the *nagual* is every bit as physically real as the person with whom it is linked".[21] The nagual, in Henne's understanding, *must* also include its physical, in-the-world, incarnation. In a different article, "A Cartography of the Uncertain: The Maya Textual Exile", he underlines "the lack of definitive divisions between self and non-self in Maya thought".[22] For Mayas, subjects do not possess a "soul" as defined in Western Christian understanding. Rather, they possess an ensemble of "spiritualities", all located in the heart. Some of these may go out for a brief stroll or can even be stolen or eaten. This ensemble of spiritualities shapes a person's character, and memory, feelings, and emotions live inside them. Spiritualities are responsible for dreams, and language begins in them as well. Furthermore, they vary from person to person, which accounts for everyone's singular behavioral

---

20 Despite the standardization of most Guatemalan Maya languages by the Academia de las Lenguas Mayas de Guatemala (ALMG; Academy of Maya Languages of Guatemala), there are still discrepancies in the spellings of certain concepts. Nagual/nawal is one of them. Most Maya languages are presently leaning towards spelling it nawal, with a w. Yet common usage still spells it as nagual. In English it has been traditionally spelled as nahual, to mark the more silent sound of the Spanish "g".
21 Henne, Nathan C.: "Untranslation: The Popol Wuj and Comparative Methodology." *The New Centennial Review* 12(2), 2012, p. 146.
22 Henne, Nathan C.: "A Cartography of the Uncertain: The Maya Textual Exile". In: Bishop, Karen Elizabeth (ed.): *Cartographies of Exile: A New Spatial Literacy*. Routledge: New York 2016, p. 28.

patterns. Usually they reside in a specific mountain on the outskirts of each town, where candles are lit for them and where they go upon the death of the body. When a body sleeps, some of these spiritualities go out to visit the outside world. It is only problematic when they do not return, and healers must be called to coax them back. We are thus looking not just at different signifiers, but at diverse ways of thinking and feeling about culture and religion. Emphasizing communal concerns, they are forms of knowledge that activate relational epistemologies, ways of knowing that muster energetic intensities, to which all Maya communities and subjects are receptive. In this way, naguales become a source of vitality, animating all subjects and allowing them to function in accordance with their nature as predetermined by their birth, according to the Maya calendar.

Thus, "The Little Boy Who Talked With Birds" can be understood based on the principle that, for the most part, Mayas see themselves dually with a part of their subjectivity in nature, expressed in their *naguales*, which constitute a diarchy in which all subjects are always and simultaneously themselves and their *nagual* at the same time, as everything in the cosmos is dual: male and female, sun and moon, morning and evening star (Venus), fire and water. This basis enables the continual regeneration of the cosmos, forever being actualized and reconfigured, one where differences are always welcome and incorporated.

The same theme reappears in "The Child Who Saw Visions". Tik-Lol, a child, had an old dog, Tusik, which barked all night long. Tik-Lol's mother tells her children that dogs who bark all night "have visions and can see strange things that other dogs cannot see" (p. 72). The child wants to know if it is true. One afternoon he wipes from the dog's eyes "the thick green ooze that gathered in the corners. Then he rubbed his own eyes with it as if it were an ointment that he had taken from Tusik's runny eyes" (p. 72). The boy then begins to have visions and see strange things. He then starts howling every night next to the dog.

Though Tik-Lol disobeys his mother, what happened to him was not punishment. The fact that human and dog subjects are alike underlines the story. They can share visions, or they do not; however, the difference is not located in their being a distinct species, but in sharing non-rational abilities to see visions. Both can share beneficiary or malicious traits, for all beingness is subordinated to cosmic equilibrium. For Mayas, there is neither an all-good God nor a humanity eternally sinning. Rather, there is a belief in the ubiquity of their deities, and in the ambiguity of their surroundings, where reciprocity with the supernatural world is the inescapable ritual gesture.

It is a cosmocentric world that does not differentiate between humanity or nature, a dynamic flow of reality for which the rivers on Earth and the cosmic river (the Milky Way) are apt metaphors, in an organic and interdependent world in which all things in the world have their own life and cosmic energy. As Mayas say, everything has *winaqil* or is *winaq*. *Winaq* is the most commonplace word in modern K'iche' for "person/ people". Thus, everything—from Earth beings like volcanoes, rivers or caves, to written words—has life, an image, a heart. These are *not* just human traits but they are shared with everything in the cosmos. In consequence, intelligence is not rational but emotional, as we see in both stories. In "The Little Boy Who Talked With Birds", the father is rational. He understands his son's behavior based on hierarchy and a sense of respect, or lack thereof. The son, on the other hand, merely enjoys his exchanges with the birds. His knowledge resides in his heart and not in the brain. In "The Child Who Saw Visions", Tik-Lol will suffer for his doings, unlike the child who spoke with birds. In both cases, however, formal logic has no place. In a holistic understanding of the cosmos, individual actions may lead to good things, or not. There is no cause-and-effect relationship. Yet in the large scheme of things, emotions restore reciprocity, balance, cyclical movement, inversion, and complementarity, as we see at the end of "The Little Boy Who Talked With Birds". In both stories there is also a world in which silence is more metaphysical than any verbal, human utterance. In both, the text expands the domain of subjectivity beyond conventional Eurocentric conceptualizations.

Where dialogue happens, trouble may follow. In "Advice From A Jackass", an ox is being exploited in the fields by a mean peasant. As they pass a jackass, the animal tells the ox not to work any longer, but "Fall down as if you are sick, or kick the executioner who drives you" (p. 60). The ox likes the idea and throws himself to the ground. Unable to make him stand upright after trying all kinds of ways to lift the ox, the furious peasant yells, "Bring me the jackass. The jackass will finish the work" (p. 70). Thus, while the jackass is whipped and has to finish the work, the ox "chewed his cud, observing from a distance his friend who had given him such magnificent advice" (p. 61). The suffering donkey lives to regret having opened his mouth.

An analogous positionality emerges in "The Curious Mice". Two country mice are watching from a roof as a woman is about to have a baby. The male mouse says that women cover the food to hide everything, whereas men are careless and hide nothing. So, if the woman has a girl, they are lost, but if she has a boy, they will be singing. The female mouse replies that, because the men lay out the traps and women are afraid of them, they can sneak into a kitchen where a woman is. For this reason, the female mouse argues that, if a girl is born, they can sing. While they bicker, a hungry cat appears, "putting to flight the two rodents" (p. 75). Fleeing, they never know what happened. As it turns out, the woman has twins, a boy and a girl, making the mice's argument moot.

In both stories, once more, dialogue, emerging from rational conceptualization, leads to misunderstandings, misconceptions, or equivocal conclusions. In the jackass's case, he pays for his good deed by becoming enslaved. At the conclusion of the mice's story, their banal argument leads them to nothing, and they never discover the outcome of their discussion. In both, there is a refusal to imagine human and animal subjectivities as separate from each other.

My last example will be "How the Serpent Was Born". The fable begins with the following statement:

> The care and devotion of a mother for her growing children is enormous. She denies herself and she pours forth the treasure of love from her heart in caring for her child. [...] A mother is a treasure. A mother is a special being whom we ought to love every moment of our lives. But many of us do not have hearts big enough to repay her for all that we make her suffer. (p. 39)

The opening rhetorical tone conditions the developments of the fable, which becomes a cautionary tale about not dishonoring your mother. The story begins when a "certain mother *wanted* to visit her son's house and *rest in the shade of his roof*. Since he was her son, he might even give her some tortillas to quiet the *great hunger* raging in her stomach" (p. 39; my italics). The words in italics mark the mother's will to be with her grown son and her status as a poor, hungry woman. The story, however, tells readers that this was not to be. When the son sees his mother approaching, he curses and tells his wife to hide their food and not feed her. To add insult to injury, when she reaches the house door, her son asks her:

> "Old woman, why do you come to my house?"
> His mother answered, "Son, I only come to rest in the shade of your roof."
> "Well, I don't believe I have anything to give you, and besides these visits bore me."
> The son and his wife had to work hard to fight the appetites that made them want to devour the succulent chicken soup right in front of the old woman who would then want a share. The old woman grew tired of sitting on the doorstep with not a kind word from her son. She turned back towards her little house, saddened by the ingratitude and indifference of that self-centered and ungrateful son. (pp. 39–40)

The overtly abusive tone would feel unreal for a Maya community, where respect for the elders and their care have been a constant for millennia. Contravening this taboo, the fable marks a rhetorical space where, though never spelledout, we implicitly have a Westernized Maya couple prioritizing individual material benefits over the community's ethics. The couple evokes Montejo's affirmation in the author's preface, mentioned at the beginning of this section of the chapter, that a fast-paced Westernization process in Guatemala has led to a loss of Maya heritage. In the fable, this information is contextual, rather than textual, nevertheless, it marks how coloniality has reorganized Indigenous subjectivities, even if they remain separated from the hegemonic Mestizo population by the caste-like social divisions that keep both communities apart. This colonialized legacy also imposes a transgressive mode of kinship that collapses Maya cosmological principles. As Henne explains in "A Cartography of the Uncertain", "the individual person as an independent entity does not exist separately from the communities in which she moves. These communities include, but are not limited to, people, food, animals, plants, and the dead" (p. 38). Subjects as individual beings are not denoted in this fashion in Maya languages.[23] The uncanny behavior of this couple denotes an exile of the self, and most likely, it also connotes a problematic relationship with other Maya realms marked by place as well.

After the mother leaves, though, we find an ironic twist in the fable. The wife brings the pot of food back out and places it on the table. When she lifts the lid, "instead of the chicken soup, she saw a poisonous serpent, coiled in the pot, its head poised, ready to strike. They want to kill it, but the snake, shaking its rattles, slithered out to hide" (p. 40). The story thus ends with the downfall of the wicked son who scorned his mother. We must note that, more important than the role of the son, wife, or mother, is the role of the snake as an executor of moral justice. The snake scares the couple. Simultaneously a symbol of justice, the snake plays out its double role in the Indigenous world, one associated with violence and revenge but at the same time revered, for it is considered sacred, as will be explained in the next paragraph. Either way, in Montejo's fable the snake confronts the couple's lack of morality and breakdown of communal traditions. In the fable, the snake plays the role of enforcer of a rough type of justice, a messenger of cosmic deities punishing those who break laws or violate taboos. At the same time, the snake condemns the breakdown of respect for the mother-figure on the part of the couple. In most cultures, a son would be expected to embrace his mother and respect her.

In Mesoamerican cultures, snakes are associated with divinity, rebirth, and spiritual power and are often looked upon with both fear and awe. In Maya and Mesoamerican cosmovision, there existed the veneration of a Quetzalcoatl-like feathered serpent that represented the conduit between the supernatural and human worlds. PreHispanic Mayas appeared as "a snake-headed cord that emerged from the belly of the Maize God and the sacred place they called

---

23 Henne argues in the same article that "the fact that indigenous languages do not have a verb for the 'be' that means 'exist' in Indo-European languages—at least not one that functions structurally like this verb—reflects this refusal to make individual existence concrete" (p. 38).

Na-Ho-Chan. Classic Kings carried it in their arms in the form of the Double-headed Serpent Bar".[24] Teotihuacan, Mesoamerica's largest city in the late preClassic and Classical period (200 BCE – 700 CE) featured the Temple of the Feathered Serpent, built around 1–150CE, which had a long underground water tunnel that also represented the conduit between the supernatural and human as previously indicated. Several feathered serpent representations appeared on the building, including full-body profiles and feathered serpent heads. The Temple of Inscriptions in Palenque, where the tomb of King Pakal was found, also had a spring to allow water to flow under the pyramid. Scholars believe the tunnel was built to allow Pakal's spirit to enter the underworld in the water that passed beneath his funeral chamber. Archaeologists working in Chich'en Itza have also found that the Temple of Kukulkan, the Feathered Serpent in Cho'ol Maya, also known as El Castillo, was built over a network of underground rivers that connect throughout the region. We also know that during the spring and fall equinoxes the shadow cast by the angle of the sun and edges of the nine steps of the Temple of Kukulkan, combined with the northern stairway and the stone serpent head carvings, create the illusion of a massive serpent descending the pyramid. Thousands of people still gather to witness the phenomenon to this day.

Indeed, in the Maya region, the Post-Classic period (700–900 CE) led to the cult of the Feathered Serpent, in the Maya region which most ethnohistorians believe originated in Teotihuacan. Ruud van Akkeren has described a Feathered Serpent cult that spread in the post-classic period from Yucatan down to the highlands of Guatemala. In the great ballcourt complex of Chich'en Itza, there is a bas relief of the high priest of the Feathered Serpent. In his understanding, the K'iche brought the cult of the Feathered Serpent into the Guatemalan highlands. Thus, considering this genealogy, we can clearly see how the serpent in Montejo's tale is not simply one connoting a facile symbol of morality and ethics. The snake is also the presence that connects the fable to Maya cosmovision and also condemns the breaking down of a Maya way of thinking. There are layers of meaning in the symbol deployed by the author. Montejo builds meaning on distinct levels, as Henne explains the *Popol Wuj* does in "Untranslation": so that each level "dialogues with and complicates a simple interpretation of an occurrence in the world that refuses to be adequately represented by factual language" (p. 132). Montejo points in a subtle way in the direction of the phantasm of

---

24 Friedel, David: "Centering the World: The Navel of the World". In: David Friedel, Linda Schele, Joy Parker (eds.): *Maya Cosmos: Three Thousand Years on the Shaman's Path*. William Morrow: New York 1993, p. 128.

Kukulkan when deploying his rattlesnake. An appeal to the rattlesnake's body summons the deity.

At the same time, the mother is not just a mother either. She may also be Ixkik', Young Blood Moon, a moon deity, from the *Popol Wuj*. She is the daughter of Kuchuma Kik', a Xib'alb'a (underworld) lord who became the mother of the Hero Twins, Junajpu and Xb'alanke.[25] In *Xib'alb'a y el nacimiento del nuevo sol*, Van Akkeren notes that Ixkik' is the preHispanic Mother Goddess. Identified with the phases of the moon, she is related to lunar eclipses, fertility, and childbirth.[26] Were we to allegorize the fable and read it as a layered duality, we would have a modern Ixkik' rejected by her Westernizing son and wife. In turn, Kukulkan, the Feathered Serpent, reminds the couple of their Maya roots, identity, and sense of community that underscores an ethical belonging that includes a holistic spiritual practice in which everything—politics, gender, aesthetics, philosophy—is contained. What happens to the couple in this fable is what Henne explained as "an exile of self related to the perception of one's relationship with other realms marked by place" (2016, p. 29).

The phantasmatic evocation of pre-Hispanic cosmological motifs gives rise to a textual interplay between the classical past and the present; indeed, it is a desire to underscore the uninterrupted continuity of Maya culture and community for more than 2,500 years. Gloria Chacón has labeled this double gaze as *kab'awil*.[27] Represented as a double-headed eagle called *k'ot*, meaning "one head looking at the sky and the other looking at the earth" by K'iche' communities, it is presumed to symbolize the classical deity Kab'awil, meaning "double-sighted deity", implying two faces, two forms, two opposing energies. The motif originated as the ancient Maya sky saurian representing the ecliptic (the path of the sun and planets through the sky) and the midday sun at the "Heart of Sky" portal. Kab'awil stands for Chacón as an ontology and hermeneutics to discuss contemporary indigenous discursive practices on a permanent double dimensionality: the preHispanic past and the ongoing Eurocentric globalized present.

---

25 As young boys descend to Xib'alb'a, outwit the nine lords who reign there, and destroy them. Afterward, they rescue the head of their father Jun Junajpu —previously killed by the lords— and transform him into a head of maize. They then ascend to the night sky as the sun and the moon, thus preparing earthly conditions for the appearance of human beings, those made of maize.

26 Van Akkeren, Ruud: *Xib'alb'a y el nacimiento del nuevo sol*. Piedra Santa: Guatemala 2012, p. 155.

27 Chacón, Gloria: *Cosmolectics and the Rise of Contemporary Maya and Zapotec Literatures*. University of North Carolina Press: Chapel Hill 2018.

Kab'awil is an episteme of the non-presence of the pre-Hispanic past, while pondering on the abject presence of Eurocentric modernity. Two kinds of thinking co-exist. One takes issue with the other, testing the "skin knowledge", the felt imprint of the present, and contrasting it with ontologies lived in a faraway past, without being able to appeal to any empirical experience of that other lost world. Yet the confrontation of a present in which the subject's body is abjectly racialized, contains the phantasmatic traces that invoke the sensuous interrelationship of body-mind-environment with that classical Mesoamerican past. A strictly contemporary Western-centric stand could never display it because the trauma of the abjection of the racialized present is not an opposite that can be reconciled. Thus, Kab'awil becomes a meditation on the self as other and explains the undecidability of words as well as their unresolvable contradictions. They are naming two realities, one in their own language and other in the language of their masters.

Final Thoughts

The logic described in the last paragraph of the analysis of Montejo's oeuvre should suffice for an explanation as to why contemporary Animal and Plant Studies theories emerging in the Global North cannot account for Indigenous representations, despite Indigenous communities' affective and transformational mode of knowing animals. Eurocentric Animal and Plant Studies are at once too capacious and too narrow. Failing to capture the very real and meaningful linkages between human and animal subjects that promote self-knowledge and communal belongingness, they also elide critical differences in the ways in which subalternized and racialized peoples of the Global South conceive animal/human relationships. They could not account, for example, for the lack of definitive divisions between self and non-self in Maya thought. As with many Indigenous ontologies, the latter also implies somatic and bodily knowing, as opposed to Eurocentric's strictly cerebral way of knowing. Whereas the latter are more anthropocentric, Mayas are more anthropocosmic. Maya subjectivity implies bodily connections to the world of animals, plants, and Earth beings, where these last aspects constitute a cosmological link. They all operate as reciprocal agents and all embody ways of knowing. Subjectivity is a blending of these relationships, and a being creatively focused on these factors. Though fraught, the relationship between human and animal subjects—to which we could add plants or Earth beings—continues to be rich with scholarly promise given human beings' urgent need for sustainable bioregions, if these issues are incorporated within globalized complexities that transcend the club-like atmosphere of Anglocentric academic debates.

Bibliography

Battiste, Marie and James (Sa'ke'j) Youngblood Henderson: *Protecting Indigenous Knowledge and Heritage: A Global Challenge*. Purich Publishing: Saskatoon, Canada 2000.

Beaulieu, Alain: "The Status of Animality in Deleuze's Thought". *Journal for Critical Animal Studies* 9(1), 2011, pp. 69–88.

Beck, Ulrich: Camiller, Patrick (trans.): "The Truth Of Others: A Cosmopolitan Approach". *Common Knowledge* 10(3), 2004, pp. 430–449.

Belcourt, Billy-Ray: "Animal Bodies, Colonial Subjects: (Re)Locating Animality in Decolonial Thought". *Societies* 5(1), 2015, pp. 1–11.

Bray, Tamara L.: "An Archaeological Perspective on the Andean Concept of *Camaquen*: Thinking Through Late PreColumbian *Ofrendas* and *Huacas*". Cambridge Archaeological Journal 19(3), 2009, pp. 357–366.

Chacón, Gloria: *Cosmolectics and the Rise of Contemporary Maya and Zapotec Literatures*. U of North Carolina P: Chapel Hill 2018.

Cummins, Thomas B. F.: "Inka Art". In: Shimada, Izumi (ed.): *The Inka empire: A Multidisciplinary Approach*. U of Texas P: Austin 2015, pp. 165–196.

Deleuze, Gilles and Felix Guattari: *A Thousand Plateaus*. U of Minnesota P: Minneapolis 1987.

Friedel, David: "Centering the World: The Navel of the World". In: Friedel, David; Chele, Linda; Parker, Joy (eds.): *Maya Cosmos: Three Thousand Years on the Shaman's Path*. William Morrow: New York 1993.

Goldberg, David Theo: *The Racial State*. Blackwell: Oxford 2002.

Grim, John: "Knowing and Being Known by Animals: Indigenous Perspectives on Personhood". In: Waldau, Paul / Patton, Kimberly (eds.): *A Communication of Subjects: Animals in Religion, Science, and Ethics*. Columbia U P: New York 2006, pp. 373–390.

Henne, Nathan C.: "Untranslation: The *Popol Wuj* and Comparative Methodology". *The New Centennial Review* 12(2), 2012, pp. 107–149.

Henne, Nathan C.: "A Cartography of the Uncertain: The Maya Textual Exile". In: Bishop, Karen Elizabeth (ed.): *Cartographies of Exile: A New Spatial Literacy*. Routledge: New York 2016.

Hirschkop, Ken: *Mikhail Bakhtin. An Aesthetic for Democracy*. Oxford University Press: Oxford 2000, p. 67.

Latour, Bruno: "Whose Cosmos, Which Cosmopolitics?". *Common Knowledge* 10(3), 2004, pp. 450–462.

Lattas, Andrew: "Primitivism in Deleuze and Guattari's *A Thousand Plateaus*". *Social Analysis: The International Journal of Social and Cultural Practice* 30, 1991, pp. 98–115.

Legrás, Horacio: "The Rule of Impurity: Decolonial Theory and the Question of Literature". In: Ramos Juan G. / Daly, Tara (eds.): *Decolonial Approaches to Latin American Literatures and Cultures*. Palgrave Macmillan: New York: 2016, pp. 19–36.

Mignolo, Walter D.: *Local Histories/Global Designs: Coloniality, Subaltern Knowledges, and Border Thinking*. Princeton U P: Princeton 2000.

Montejo, Victor: Kaufman, Wallace (trans.): *The Bird Who Cleans the World and other Mayan Fables*. Curbstone: Willimantic, CT 1991.

Nocella II, Anthony J. / Sorenson, John / Socha, Kim / Matsuoka, Atsuko: *Defining Critical Animal Studies: An Intersectional Social Justice Approach for Liberation*. Peter Lang: New York 2014.

Quilter, Jeffrey: *The Ancient Central Andes*. Routledge: New York 2014.

Salomon, Frank: Salomon, Frank / Urioste, George L. (trans.): "Introductory Essay". *The Huarochiri Manuscript: A Testament of Ancient and Colonial Andean Religion*. U of Texas P: Austin 2001.

Spivak, Gayatri Chakravorty: "Can the Subaltern Speak?". In: Nelson, C. / Grossberg, L. (eds.): *Marxism and the Interpretation of Culture*. Macmillan: Basingstoke 1988, pp. 271–313.

Stengers, Isabelle: Bononno, Robert (trans.): *Cosmopolitics I*. U of Minnesota P: Minneapolis 2010.

Stengers, Isabelle: Bononno, Robert (trans.): *Cosmopolitics II*. U of Minnesota P: Minneapolis 2011.

Van Akkeren, Ruud: *Xib'alb'a y el nacimiento del nuevo sol*. Piedra Santa: Guatemala 2012.

Viveiros de Castro, Eduardo: "Cosmological deixis and Amerindian perspectivism". *Journal of the Royal Anthropological Institute* (N.S.) 4(3), 1998, pp. 469–488.

Watson, Matthew C.: "Cosmopolitics and the Subaltern: Problematizing Latour's Idea of the Commons". *Theory, Culture and Society* 28(3), 2011, pp. 55–79.

Watson, Matthew C.: "Derrida, Stengers, Latour, and Subalternist Cosmopolitics". *Theory, Culture & Society* 31(1), 2014, pp. 75–98.

Beatriz Rivera-Barnes
Sadder Tropics: The Hate of Nature in Juan José Saer's *El entenado* and Dorian Fernández-Moris's film *Desaparecer*

Abstract: Why refer to the hate of nature? Why read hate into Saer's narrative and into a 2015 Peruvian film with relatively negative reviews? "Do we need hate?" the ethnolinguist James Underhill asks before coming to the conclusion that hate can no longer be simply denounced and stored away in a category for negative emotions, that it must be engaged as a concept. "I hate nature" is the title of a short piece by Dr. Martha Schwartz, a professor of Landscape Architecture at Harvard University. "I hate spiders. I hate earthquakes. I hate genetic diseases and AIDS. Nature is dirty". Hating nature can also evoke the destruction of the air, oceans, rivers, animals, trees, climate, and ozone. The two terms, hate and hatred, have numerous synonyms: animosity, aversion, repugnance, loathing, odium, execration, and anger, just to name a few. Then there are some efforts at establishing distinctions. While anger is sudden, hatred is lingering. While anger is directed at people, hatred can be felt for people, things, actions, perhaps even places. According to *The Merriam-Webster Thesaurus* repugnance applies to that which one feels when summoned to do something from which one instinctively draws back, whereas aversion is the turning away of the feeling or the mind from a person or things. The hate or the hatred of nature can entail repugnance toward an uncharted and dangerous wilderness, toward blood, relentless swamps, headless cadavers, the smell of human flesh cooking on the grill, putrefaction, filth, birth, death, or an orgy where, "no tenían en cuenta ni edad, ni sexo, ni parentesco" (Saer 2016, p. 72) ["neither age, nor sex, nor kinship mattered"]. Hatred also calls for reciprocity. Something hated will not love in return; therefore, the idea of nature hating man in return calls for some analysis and consideration. This study is an ecocritical approach to a novel and a film of the jungle, as well as to nature itself.

Keywords: Nature, Hate, Juan José Saer, Dorian Fernández-Moris

In his narrative titled *El entenado*, translated into English as *The Witness*, but more like the *Stepchild* or the *Foundling*, the Argentinian Juan José Saer (1937–2005) draws inspiration from the discovery and exploration of the Río de la Plata, a river and estuary formed by the confluence of the Uruguay and Paraná Rivers: "Ese río, que atravesaba por primera vez, y que iba a ser mi horizonte y mi hogar durante diez años, viene del norte, de la selva, y va a morir en el mar que el pobre capitán llamó dulce. Ellos lo llaman padre de ríos" (p, 39) ["That river, that I was crossing for the first time, and that would be my horizon and my home for ten years, comes from the north, from the jungle, and flows into the sea that the poor captain called sweet. They call it the father of rivers"].[1] Already, the tone is melancholic. In his memoir, Claude Lévi-Strauss referred to the tropics as *tristes tropiques*, a puzzling juxtaposition. The anthropologist seems to be challenging the caricature. Could Saer's narrator be another melancholic anthropologist in the making? A simpler question: why sad? It could be political and have everything to do with the intrusion of the Europeans, for the noble savage and the noble landscape have been invaded, pillaged, damaged, and conquered, yet that response would be too simple an answer to a simple question.

"What kind of land is this?" Cabeza de Vaca asked himself in the middle of a swampland. A decade before, foundering in the Florida Everglades, Juan Díaz de Solís set out in search of the passage between the Atlantic (Mar del Norte) and Pacific (Mar del Sur) Oceans. In early 1516, Solís reached what he called Mar Dulce (Sweet-Water Sea), the estuary that would later be called Río de la Plata. In the preface to *The Improbable Conquest, Sixteenth Century Letters from the Río de la Plata*, the editors García Loaeza and Garrett describe how this brief exploration of this "sweet-water sea ended when the natives attacked, killed and allegedly devoured a landing party led by Díaz de Solís" (p. 30).[2] Díaz de Solís was among the members of the party killed by the natives. Is this where hate and the promise of repugnance and eternal sadness begin?

The sudden and brutal massacre could be what rendered the conquest *improbable*, as García Loaeza and Garrett suggest with their title. The members of the party were killed, so they did not conquer. Something improbable is not likely to be true, not likely to happen. Was there in fact no conquest? In the epilogue, García Loaeza and Garrett explain that the idiosyncrasies of the conquest underscore that it is misleading to speak of the conquest of America because there were many conquests whose outcomes were determined by myriad factors such as distance, weather, politics, and culture (p. 104). The implication could be that there was an accidental element to a conquest, if conquest there was, and that the accident and its outcome (the possible conquest) are tied either to nature (distance, weather) or to the people living in this nature (politics, culture). In

---

1 Saer, Juan José: *El Entenado*. Lanzallamas: San José, Costa Rica 2016. All translations of *El Entenado* are my own.
2 García Loaeza, Pablo, and Garrett, Victoria L. (eds.): *The Improbable Conquest*. The Pennsylvania State University Press: University Park 2015.

Image 1. Río de la Plata. https://wordlesstech.com/wp-content/uploads/2011/06/Rio-de-la-Plata-map.jpg

other words, nature and the people living in the natural world contributed to all the possible outcomes. Obviously, with nature comes the discovery of nature; the initial awe, the lack of adjectives to describe it, the taking possession, the natural disasters, suffering in the tropics, and nature as antagonist.

In *El entenado*, Juan José Saer begins to write at the precise moment that history speaks no more, the silence after the massacre. His discourse is a philosophical remaking of the discovery not only of the estuary, but of nature itself. Now to the questions: Why refer to the hate of nature? Why read hate into Saer's narrative and into a 2015 Peruvian film with relatively negative reviews?

(Pimentel).³ "Do we need hate?" the ethnolinguist James Underhill asks before coming to the conclusion that hate can no longer be simply denounced and stored away in a category for negative emotions, that it must be engaged as a concept (p. 144).⁴

"I hate nature" is the title of a short piece by Dr. Martha Schwartz a professor of Landscape Architecture at Harvard University.

> I hate spiders. I hate earthquakes. I hate genetic diseases and AIDS. Nature is dirty. Birds poop all over my car. Nature causes death. It takes up too much space. It brings ice onto the roads, germs into our living rooms, and water through the windows.
> REAL Nature is disconnected from our FANTASY about it: I, like most people, want Nature... functional and in its place.⁵

Hating nature can also evoke the destruction of the air, the oceans, the rivers, the animals, the trees, the climate, the ozone. The two terms, hate and hatred, have numerous synonyms: animosity, aversion, repugnance, loathing, odium, execration, and anger, just to name a few. Then there are some efforts at establishing distinctions. While anger is sudden, hatred is lingering. While anger is directed at people, hatred can be felt for people, things, actions, perhaps even places. According to *The Merriam Webster Thesaurus* repugnance applies to that which one feels when summoned to do something from which one instinctively draws back, whereas aversion is the turning away of the feeling or the mind from a person or things. The hate or the hatred of nature can entail repugnance toward an uncharted and dangerous wilderness, toward blood, relentless swamps, headless cadavers, the smell of human flesh cooking on the grill, putrefaction, filth, birth, death, or an orgy where, "no tenían en cuenta ni edad, ni sexo, ni parentesco" (Saer 2016, p. 72) ["neither age, nor sex, nor kinship mattered"]. Hatred also calls for reciprocity; something hated will not love in return; therefore, the idea of nature hating man in return calls for some analysis and consideration.

In *El entenado*, the protagonist's interior monologue often suggests that nature hates: silent, monotonous, odorous, and sickening. In this sense, the hate of nature is also nature's hate. "I hate travelling and explorers", thus begins Claude Lévi-Strauss's memoir *Tristes Tropiques*.⁶ Indeed, the first question that comes to mind

---

3 Pimentel, Sebastián: "*Desaparecer*, una película sin fuerza ni personalidad propia". *El Comercio*, 05/18/2015, from https://elcomercio.pe/luces/cine/desaparecer-pelicula-fuerza-personalidad-propia-363719.
4 Underhill, James W.: *Ethnolinguistics and Cultural Concepts*. Cambridge University Press: London 2012.
5 Schwartz, Martha:http://www.marthaschwartz.com/academic/writings/i-hate-nature/
6 Lévi-Strauss, Claude: *Tristes Tropiques*. Penguin: New York 1973.

is why these tropics are *tristes*, sad. Is it simply a good alliterative title? Obviously not, as there is something inherently sad about the tropics. For fifteen years, Lévi-Strauss planned to write this memoir but *shame* and *repugnance* prevented him from doing so. The anthropologist wondered if it was worth his while taking up his pen to perpetuate sickening boredom and useless shreds of memory (p. 17).

While traveling through the Peruvian Amazon, the filmmaker Dorian Fernández-Moris also discovered these elements of hatred that would lead to the making of *Desaparecer*. In an interview with Jorge Schwartz, he describes how he heard about rosewood and its exploitation,

> que incluso mafias habían cometido acciones de lesa humanidad en la Amazonía, y lo que me decían los pobladores era que no había forma de que esto se sepa afuera, a nadie le importaba, me decían: parece que nosotros estamos abandonados en el mundo y más allá de Iquitos a nadie le interesa lo que sucede aquí.
>
> [that even mafias had committed crimes against humanity in the Amazon whose inhabitants told me that there was no way for that to be known by the outside world because no one cared: it seems that we have been abandoned by the outside world and no one is interested in what happens in Iquitos].

Unlike novels such as *El viejo que leía novelas de amor* that appear tailor-made for an ecocritical approach, Saer's novel is more likely to be considered a historical novel in that it borrows from a historical event. As for Fernández-Moris's film, it has been labeled "a thriller that raises the bar" and that "milks the jungle setting for all its worth" (Zelaya Miñano).[7] In an interview at the University of São Paulo in 1997, Saer affirmed that Latin American literature was, for him, a historical category, or rather a geographical one, not an esthetic one: "La literatura latinoamericana para mí es sólo una categoría histórica, o ni siquiera histórica, una categoría geográfica, pero no es una categoría estética" (Schwartz)[8] ["For me Latin American literature is just a historical category, or not even historical but just a geographical one, and in any case not an aesthetic category"].

Consequently *El entenado*, one of three of Saer's novels set in the past, has often been considered a historical novel by critics. It was included in Seymour Menton's 1983 publication *La nueva novela histórica de la América Latina*. Many critics, however, argue that there is no intention of historical reconstruction in *El entenado*. Florencia Abbate writes that this is not a historical novel on four

---

7 Zelaya Miñano, Ernesto: "Desaparecer, a Jungle-set Thriller that Raises the Bar". *Screenanarchy*, 05/19/2015, from http://screenanarchy.com/2015/05/review-desaparecer-a-jungle-set-thriller-that-raises-the:bar.html
8 Schwartz, Jorge. "Saer en la Universidad de São Paulo", from https://www.avizora.com/publicaciones/monosavizor/entrevista_saer.htm.

different levels: ideological and discursive improbability (no effort to imitate the thought and language of the time), the scarcity of place names, the scarcity of historical names, and little attempt to date the narrative (p. 17).[9]

Moreover, the narrator constantly reiterates his orphanhood, making history, or the past, an absolute void. Julio Premat avers that the cosmic orphanhood of the narrator creates a cultural, temporal, and geographical alterity and a general regression that will eventually call for a renaissance and the construction of affiliation (p. 51).[10] In this case, history is in the future.

In an essay titled "Viaje a la semilla", Mercedes López Baralt analyzes a series of authors such as Alejo Carpentier, Octavio Paz, and Pablo Neruda who have undertaken what she sees as a *viaje a la semilla* (journey to the seed, or origins, or the quick) with a re-writing of colonial letters, a profoundly decolonizing gesture, in her opinion (p. 36).[11] This effort can be akin to what Astvaldur Astvaldsson describes as the "necesidad de rescatar una literatura ausente, la memoria de textos borrados, destruidos aún antes de que fueran escritos y en los cuales están inscritas las prefiguraciones del porvenir" (p. 68) ["the necessity of rescuing an absent literature, the memory of erased texts destroyed even before they were written, and in which are inscribed the prefiguration of the future"].[12]

Beatriz Sarlo, in turn, also affirms this novel does not respond to what is called a historical novel today, and that it is, instead, a philosophical fable (p. 314).[13] Florencia Abbate suggests that *El entenado* is a philosophical fable as well and likens it to Joseph Conrad's *Heart of Darkness*, where the narrator evokes the past, his encounter with nature, his interminable journey to the heart of nature that rendered a return to the city quasi impossible. This is the exact opposite of Robinson Crusoe's trajectory, for while Crusoe recreates civilization in the heart of nature, Conrad's Marlow and Saer's narrators see civilization as a result of their encounter with nature, with the other (p. 21). Such a journey strongly

---

9   Abbate, Florencia: *El espesor del presente*. Editorial Universitaria Villa María: Córdoba, Argentina 2014.
10 Premat, Julio: *La dicha de Saturno*. Beatriz Viterbo: Rosario, Argentina 2002.
11 López Baralt, Mercedes: "Viaje a la semilla: la reescritura contemporánea de la letras coloniales." In: Chocano / Rowe (eds.): *Huellas del mito prehispánico en la literatura latinoamericana*. Iberoamericana Vervuert: Frankfurt 2011, pp.23–39.
12 Astvaldsson, Astvaldur: "Mitos, paisaje, y modernidad en la literatura latinoamericana". In: Chocano / Rowe (eds.): *Huellas del mito prehispánico en la literatura latinoamericana*. Iberoamericana Vervuert: Frankfurt 2011, pp. 67–90.
13 Sarlo, Beatriz. *Escritos sobre literatura argentina*. Buenos Aires: Siglo XXI Editores Argentina, 2007.

suggests the geographical category, perhaps even the landscape as genre. *El entenado* therefore warrants an eco-critical approach, as does *Desaparecer*, in which the jungle is protagonist, antagonist, victim, and perpetrator. Both the narrative and the film describe what Astvaldsson would consider to be "una manera particular de relacionarse con el paisaje que nos circunda y, en fin, con la comunidad y la propia existencia humana" (p. 69) ["a particular way of relating to the landscape that surrounds us and, finally, to the community and human existence itself"].

The opening pages of *El entenado* are reminiscent of a picaresque novel and also of Melville's *Moby Dick*. The narrator, who will always remain anonymous, immediately reveals that he is now an old man who spends his time in the cities because "en ellas la vida es horizontal" (p. 11) ["because life is horizontal there"]. It was not always so. Being an orphan made the protagonist seek the ports, merchants, prostitutes, and alcohol, until the day he had his first prostitute and his first alcoholic beverage and became a man. From foundling, or *niño hallado*, to adult, a sudden metamorphosis and perhaps the first renaissance, the alcohol and the prostitutes are thus instrumental in the construction of very first affiliation.

From that moment on, in search of another possible affiliation, the ports ceased to be enough, and the narrator immediately felt compelled, if not driven, to reach the other side, the opposite shore where fruit had to be more delicious and more real, the sun more yellow and benevolent, and human deeds more comprehensible, just, and defined (p. 12).

In the first scene of *Desaparecer*, a little girl in a jungle village walking home from school is lured by a faceless man who asks her if she likes dolls. The end of this scene depicts the doll floating in the shallows of the brown river. A young girl's shrill scream can be heard. The second scene takes the audience to a safe, modern world, the upscale neighborhood of Miraflores in Lima. Although Lima is not exactly a safe city, from the onset a contrast exists, safe city versus dangerous jungle. Somewhere upstairs in one of those plush Miraflores buildings, a couple is celebrating an impromptu reunion and a birthday. Immediately, one senses something pleading about Giovanni (in a flashback that will come in a subsequent scene, Giovanni urges Milena not to return to the jungle and to stay in Lima where it is safe, but the audience is not yet aware of Milena's relationship with the jungle). For the time being, perhaps Giovanni's malaise has everything to do with his intention to propose to Milena. Soon enough, however, inklings of the conflict surface. Milena, an ecologist and activist working in a village on the banks of the Amazon River, assures Giovanni that this will be her last trip to the jungle. He insists that she is through with her work in the jungle. Why return?

Milena's attitude suggests that she has unfinished business in the jungle and must therefore return.

This yearning for the unknown, or rather the half-known, "el lugar perfecto para hacer ondular deseo y alucinación" (Saer 2016, 13) ["the perfect place to undulate desire and hallucination"], could easily describe Milena's feelings. They make the narrator of *El entenado* join an expedition to the Indies. This is the beginning of the sixteenth century, the age of exploration when the newly discovered Americas still need to be invented, when nature and the natural, so absent in Europe, had yet to be discovered.

In 2015, Milena sets out for the known jungle, charted territory, and disappears; the jungle has swallowed her. She is not the first to disappear; several children from the jungle village have recently vanished, to the villagers' rage that is unfortunately turned inward, toward nature itself, toward the river, their surroundings, the fruit of the jungle, the Yakuranas, the mythical fish men of the Amazon.

The narrator of *El entenado* describes three months of monotonous blue, unlimited sky, empty expanse, dilated blue, and feeling small, like ants in the middle of a desert (pp. 11–15) (this is reminiscent of Lévi-Strauss's "sickening boredom"). Just as Columbus used the colors green and brown to describe the American landscape, for lack of other descriptive adjectives, the narrator continually uses the colors blue and yellow. Living in this constant landscape of boredom, the narrator reiterates that he was an adolescent, that his shape was juvenile, his virility incomplete, that the absence of women made the honest fathers and husbands who surrounded him forget what would otherwise have been repugnant on land; the act became each day more *natural* (p. 16). Instead of hating his abusers, the orphaned narrator ended up seeing them as father figures. This is perhaps a second renaissance, yet another construction of affiliation. There is enough material here for an entire narrative: a young orphan in search of the great beyond goes on an expedition to the Indies. The absence of women and the cabin boy's youthful appearance make the members of the crew forget what they would have otherwise abhorred on land. The abusers are like father figures. The abuse is described as pleasant and *natural*. Up until they sight land, the orphan is a foundling, safe, bored, adopted.

Upon disembarking, "nos dispersamos como animales en estampida" (p. 18) ["we dispersed like animals stampeding"]. The land was for the taking, the grabbing, the pillaging. While some members of the crew ran aimlessly in all directions, others jumped, others lit fires, others made fun of a bird, others chased a bird, others climbed trees, others dug the earth. The next day, "El sol teñía de rojo el mar y ennegrecía las siluetas de los barcos" (p. 19) ["The sun dyed the water red and blackened the silhouette of the boats"]. This sentence definitely casts a pall, red and black, over the nefarious unknown that lies ahead. Immediately, the captain decides to sail south, positing that these lands were not the Indies, merely an unknown world.

From monotony to silence, the point of departure becomes an unloved place, reminiscent of Cabeza de Vaca's "land so strange", so often described as *poor*. The reader of the *Naufragios* could surmise that the land is poor in terms of agriculture since the explorers find very little growing in the marshes. It is, however, also poor in terms of cities, and in terms of souls. "But Cabeza de Vaca also calls this land *mala*, bad, evil, and the people wretched, an impossible land, impossible to dwell in it, and impossible to escape from it" (Rivera-Barnes, p. 212).[14]

From the moment they first saw land, the attitude of the captain in Saer's narrative changed, and he seemed to have retreated into another world. His gaze, inalterable and dignified, "iba fija en los árboles que crecían al pie de la loma, donde terminaba la playa y comenzaba la selva" (p. 21) ["was fixated on the trees growing at the foot of the hill where the beach ended and the jungle began"]. Without uttering a single word, the captain continued staring into the distance. The only sounds were those of the waves breaking against the sands. Five minutes later, he let out a sigh. Sixty years have elapsed since the narrator heard this sigh, but it left such an impression on him that he feels the memory will accompany him to his dying day. The other members of the crew, in turn, felt panic. The captain finally turned around and walked back to the boat. They proceeded to sail down the coast.

They sailed until they reached what the narrator described as brown and fresh waters, the mouth of a river or an estuary that would later be baptized Río de la Plata by the Italian Sabastián Gabto, and that the captain of Saer's narrative dubbed the Sweet-Water Sea and claimed it for the King of Spain with mechanical gestures. Each time they disembark, the narrator compares the crew to a colony of ants coming from nothingness. He also refers to the river as a savage one, and its odor to be one of primeval beginnings, of humidity and growth: "Salir del mar monótono y penetrar en ellos fue como bajar del limbo a la tierra. Casi nos parecía ver la vida rehaciéndose del musgo en putrefacción" (p. 27) ["Leaving the monotonous sea and entering the rivers was like going from limbo to earth. We felt that life was reborn from the putrefying mosses"]. The narrator adds that

---

14 Rivera-Barnes, Beatriz: "Is there such a Thing as too much Water?". In: Stavans, Ilán (ed.): *Chronicle of the Narváez Expedition*. WW Norton & Co: New York 2012, pp. 204–217.

Image 2. According to the International Hydrographic Organization based in Monaco, the 219-kilometer-long Río de la Plata separating Argentina from Uruguay begins at the mouths of the Paraná and the Uruguay Rivers. At times, it is considered a gulf or a marginal sea. It is also held to be an estuary. https://upload.wikimedia.org/wikipedia/commons/0/0b/Rio_de_la_Plata_BA_2.JPG

the absence of humans augmented the feeling of primeval beginnings, and he repeats the word *primeval* again when referring to the river banks.

"Después del bautismo y de la apropiación, esa tierra muda persistía en no dejar entrever ningún signo, en no mandar ninguna señal" (p. 26) ["After the baptism and the appropriation, this silent land refused to cede any signal or to send any sign"]. In other words, the land may have been appropriated, or so the conquerors believed, but the land refused to yield and it remained a monolithic block of sorts. The silence could very well be a product of a mutual lack of understanding. All suggest an improbable conquest.

"Tierra es esta sin…" (p. 31) ["Land this is without…"]. Those were the captain's last words, the ones he uttered before an arrow went through his throat. Without what? The reader will never know what the land lacked. Not only are the tropics sad and boring, but now they lack something. Soon, the narrator realizes that every member of the crew is dead and that, "con la muerte de esos hombres que habían participado en la expedición, la certidumbre de una experiencia común desaparecía y yo me quedaba solo en el mundo" (p. 31) ["with the death of those men who had participated in the expedition, the sureness of a common experience disappeared and I remained alone in the world"]. Once again, the impossibility of history.

These are the short moments when history and the narrative meet. Juan Díaz de Solís discovered this estuary and called it The Sweet-Water Sea, just like Saer's captain. Having disembarked, the members of the crew were immediately attacked and killed by a group of indigenous peoples, either the Charrúas or the Guaraní. In *El entenado*, this tribe goes by the name colastinés. Only a fourteen-year-old cabin boy by the name of Francisco del Puerto survived: the unnamed protagonist, the eternal add-on, adoptee, foundling of the narrative that the critic Ravetti deems to be either an anthropological proposal or a speculative anthropology (p. 386).[15]

In her essay on cannibalism, Ravetti analyzes various myths having to do with devouring the other that sustained cannibalistic practices. In this effort, Ravetti makes reference to Eduardo Viveiros de Castro's text *O nativo relativo* in which the anthropologist proposes a relativism and affirms that the truth of what is relative is the relation itself, the content and modes of exchange between the subjects (p. 386). On the one hand, there is the anthropologist, on the other, the native. On the one hand, there is the experience of the anthropologist, and on the other, the experiment that Viveiros de Castro considers to be a fiction controlled by the experience. In other words, the fiction is anthropological, but its anthropology is not fictional, Viveiros de Castro affirms before asking what this fiction consists of. Ravetti, in turn, finds a convergence between the issues raised by Viveiros de Castro and Saer's speculative anthropology.

As a result, Ravetti's hypothesis is that we can recreate a system of knowledge with Saer's *El entenado* as its point of departure. It is in this zone, the Río de la Plata region, that we find myths underpinning the practice of cannibalism, "esta novela establece un nuevo paradigma de tradición, al recrear un tribu que realiza anualmente un festín orgiástico con carne humana que, como veremos, tiene el objetivo de recordar y celebrar la condición humana de los integrantes de la comunidad" (p. 388) ["this novel establishes a new paradigm of tradition by recreating a tribe that holds a yearly orgiastic feast with human flesh so as to remember and celebrate the human condition of the members of the community"].

---

15 Ravetti, Graciela: "Los mitos guaraníes sobre el canibalismo y su relación con *El entenado*." In: Chocano / Rowe (eds.): *Huellas del mito prehispánico en la literatura latinoamericana*. Iberoamericana Vervuert: Frankfurt 2011, p.386–394.

Florencia Abbate explores the obsessive ritualistic need of this tribe to reestablish a social being on a yearly basis and comes to the conclusion that it indicates a very intense awareness of time and knowledge that their society is not even guaranteed a right to existence. "El apocalipsis no está al final de un recorrido de duración incierta, sino que es una amenaza permanente, y por eso la historia debe, cada tanto, recomenzar" (p. 20) ["The apocalypse is not at the end of a journey of an undetermined duration, it is a permanent threat, and that is why history must begin again every now and then"].

There is no occurrence of cannibalism in *Desaparecer* as there is in *El entenado*, but it is ever-present, perhaps more than in *El entenado* where it only takes place once a year, whereas in *Desaparecer* the people seem to be devouring each other on a daily basis. The children of the village of Nueva Esperanza are disappearing. The angry villagers blame the mythical fish men, the Yacurunas, suggesting that the environment itself, namely the river, is consuming the children and it is therefore a cannibalistic river. Likewise, Milena is swallowed up by the jungle river. Giovanni, in turn, has no choice but to leave the city and go to the jungle in search of Milena. Upon arrival in Iquitos, he is nearly shot to death by some locals on motorcycles who realize at the last minute that he was not the one they were supposed to kill, and they spare him. A few hours later, a friend warns Giovanni about the dangers of Iquitos and the nearby jungle that should be respected, if not feared. The friend also assures Giovanni that Milena knew the jungle well and that she loved helping the people live with their environment.

On the way to the jungle village of Nueva Esperanza via boat, the camera focuses on the logging along the riverbanks. This is the Yanayacu River, one of the many tributaries of the Amazon. When Giovanni reaches Nueva Esperanza a voice can be heard on the loudspeakers regretting so many sins and praying for God to forgive their sins and find their loved ones. This message puts forward an element of culpability in the disappearances and suggests responsibility comes from within. In *Rage and Time*, the German philosopher Peter Sloterdijk claims that the psychopolitical constellation of rage and time (or rage and history) is anticipated by the theological constellation of rage and eternity:

> The problem that exists between God and us contemporaries is not that we are too far away from him. Rather, God would get too close if we were to take his offerings seriously. No quality of the God of the theologians reveals this better than the most embarrassing among them: God's wrath. (p. 73)[16]

---

16 Sloterdijk, Peter: *Rage and Time*. Columbia University Press: New York 2012.

It could very well be that God's wrath is precisely the hatred of nature.

In the next scene, the lieutenant governor of Nueva Esperanza—the owner of the constant voice on the loudspeakers—tells Giovanni and the captain in charge of the investigation that a curse has fallen on this town, "they want everything, river, water, streams, they are stealing our sons". When the captain asks why there has been no appeal to the police, the lieutenant governor replies that, in the jungle, they take the law into their own hands because their only link with the city is the river, and that in the jungle everything is different, they go by other rules.

Indeed, the jungle goes by its own rules. It will perhaps take the narrator of *El entenado* ten years to understand these rules. As the only survivor, he is once again an orphan. It is as if he could not shed the void of the past. Every single one of the paternal figures he had for a few months has disappeared. Again, the narrator must reconstruct an affiliation. He happens to be the only member of the crew spared by the colastinés who will soon adopt him. While children are lost in the 2015 film, in 1516, children are also taken by the river, but only to be found and to be subsequently taken in. The narrator becomes the foundling of the colastinés. In this sense, the English translation of the title of the narrative is insufficient. The narrator is indeed an eternal witness, someone perpetually standing outside looking in. However, witnesses are not usually taken in to be adopted, and it is precisely the adoption by the colastinés that allows this cabin boy to become a witness.

This new state of orphanhood will be short-lived. Premat writes that the cabin boy exists in a "dinámica hiperbólica de substituciones y acumulaciones de modelos asimilados, perdidos, recuperados, elaborados: para él la cuestión de los orígenes, omnipresente y obsesiva, solo parece concernir la filiación masculina" (p. 66) ["Hyperbolic dynamic of substitutions and accumulation of assimilated models that are lost, recovered, elaborated: for him the question of his origins, omnipresent and obsessive as it is, only seems concerned with masculine filiation"].

Spared by the colastinés, the cabin boy becomes the witness. What he first witnesses after the massacre of the entire crew is his past being cooked, and then eaten. Although the colastinés treat the cabin boy with respect and give him water and fruit, they laugh at him while yelling the word/sound *def-ghi! Def-ghi!* The next day, however, the cabin boy feels that he has become used to his captors, so much so that, had it not been for their decapitated cadavers piled up on the banks of the river, his crew members could become a distant dream (p. 37–38).

This foundling, this witness, appears very adaptable. After having been made to run for miles with his captors, he begins to experience a rebirth. "Entenado y todo, yo nacía sin saberlo como el niño que sale, ensangrentado y atónito de esa noche oscura que es el vientre de su madre" (p. 42) ["The foundling that I was, I was being born without knowing it, like the child who emerges bloody and surprised from the dark night that is the belly of his mother"]. He also alludes to "el olor matricial de ese río" (p.42) ["the matricial odor of the river"] and again to a mute and desert land when he gazes at the decapitated cadavers being put on the grills (p. 46).

Instead of repulsion at the sight of the human grease dripping on the embers, the narrator finds the smell of human flesh cooking to be pleasant (p. 54). He wonders if this is due to his hunger, or to the fact that there was about to be a feast and that he, the eternal "extranjero, no quería quedar afuera, me vino, durante unos momentos, el deseo, que no se cumplió, de conocer el gusto real de ese animal desconocido" (p.54) ["eternal outsider, I did not want to be left outside, and I experienced for a few moments the desire, left unsatisfied, to taste that unknown animal"]. He even added that the smell made his mouth water and described the attraction and repulsion at the sight of people chewing human flesh under the intense heat and blinding sunlight.

This was not a daily ritual. This orgiastic feast only took place once a year in order to remember and celebrate the human condition. What is it about the human condition that they remember? Ravetti points out that trauma comes from what is forgotten (p. 392). By eating human flesh, the members of the colastiné tribe remember that there was a time when they ate each other and were auto-destructive, another manifestation of hatred. This shameful past must not be forgotten lest it gain strength and become trauma, hence, the yearly feast. In this way, they dominate the animal soul within themselves. Perhaps how they dominate nature? One of the members of the tribe tells the narrator how much they hate those that still eat each other and consider them less than human, therefore edible. Did this mean that they considered the members of the crew that they consumed cannibalistic, like members of other tribes? Whatever their judgment of the crew they consumed may be, Staden considers cannibalism to be a form of manifestation of great hatred (p. 267).[17] In this case, the hatred must be relived and remembered, and human flesh must be eaten on a yearly basis. Usually, they ate the flesh of the cannibalistic tribes they disdained because they reminded them of what they once were. Such is not the case with the European conquerors whom they did not know. However, there is an element of cannibalism in the conqueror/explorers comportment since they immediately claimed

---

17 Staden, Hans. *Viajes y cautiverio entre los caníbales*. Nova. Buenos Aires 1961.

the land for themselves, something akin to consuming it. It could be that these explorers were no different from the savage tribes considered so edible.

It only took the narrator two or three days to realize what was behind this cannibalistic rite, "Dos o tres días me habían bastado para comprobar de qué fondo negro tenían que subir esos indios tirando con fuerza hacia el aire transparente para poder mostrar, en lo externo de este mundo, un aspecto humano" (p. 79) ["Two or three days were enough for me to understand from what black abyss these Indians had to emerge with force toward the open air, in order to prove their humanity"]. The feast may also be a yearly recreation of the cycle of hate, much like the cycle of lust that Shakespeare describes in Sonnet 129. From anticipation to disgust, a feast goes through three stages, the preparation, the feast itself, and the aftermath, much like Shakespeare's sonnet:

> Enjoy'd no sooner but despised straight;
> Past reason hunted; and no sooner had,
> Past reason hated, as a swallowed bait

Moreover, inside that triptych there is yet another triptych, the feast itself, in three parts, the consumption of human flesh followed by the consumption of alcohol followed by the orgy. While eating human flesh, the colastinés become sadder and sadder. The consumption of alcohol and the brutal orgy exacerbate the sadness that overcomes them. The next day, the shame will be internalized for an entire year. While in the sixteenth century, a tribe is cannibalistic once a year, in the twenty-first, the environment is cannibalizing the children of a village almost on a daily basis.

Throughout the narrative, the words / sounds *def-ghi* are being directed at the witness. In time, the narrator comes to realize that the words have multiple meanings tied to what the tribe members expected of their foundling, something that they could not obtain with his death, rather with his constant presence (p. 86). Once again, an affiliation has been established that will only be threatened when the narrator comes across he who will replace him, the eternal return, so much so that the narrator felt he was seeing himself, but "*yo no venía en esas embarcaciones—venía, eso sí, un hombre vivo*" (p. 95) ["I was not the one arriving on those boats, but a live man indeed"]. This man, his double, the other, whom the colastinés call *def-ghi*, gazes at the narrator with intense hatred (p. 98). The cause for the hatred is not obvious. This prisoner, this other, seems to know exactly his role of *def-ghi*, but that is not reason enough for hatred toward the narrator who spent so much time delving into the meaning of his adoption by the colastinés. After the feast, the new prisoner is released, whereas the narrator remains with the tribe. Soon enough, he will realize that if they did not send him away, it was because they did not know where to send him. By then, the narrator has begun to consider this "horizonte de agua, arena, plantas y cielo" ["horizon of water, sand, plants and sky"], as something definite, "algo definitivo" (p. 105). The promise of leaving was erased with the years: "Cuando nos olvidamos es que hemos perdido, sin duda alguna, menos memoria que deseo. Nada no es connatural" (p.105) ["When we forget it is, without a doubt, that we have lost desire, rather than memory. Nothing is innate in us"].

Throughout the narrative, he who is writing reminds the reader, he who is reading, that he is writing sixty years later and that he is now an old man. Ten years into his stay in the jungle, he remembers the afternoon that the colastinés put him on a canoe and sent him back to his people. As usual, a member of the tribe laughs and utters the sound *def-ghi*, but this time the laughter is accompanied by an element of humor. Now that the narrator has been immersed in the language, he can understand his interlocutor saying and repeating as he holds his arm to his mouth and pretends to eat his own flesh, "yo soy el que, en broma, te decía que te iba a comer" (p. 174) ["I am the one who told you jokingly that I was going to eat you"]. It is interesting to note that a member of the tribe is capable of joking about himself.

There is a double or triple writing: the author Saer recreating the Latin American landscape, the old man remembering his past in the jungle, the colastinés choosing the cabin boy to be their witness. What was this tribe expecting? Were they expecting an oral rendering or a written one? To whom was the cabin boy supposed to report? To the Europeans whom they did not know? Astvaldsson writes that the more interpretations to a story make the story all the more interesting. The writing supposes reading, just as the telling supposes an ear. Astvaldsson writes that to read is to rewrite because each reading brings new meaning (p. 74). In this case, to rewrite, as Saer and his narrator do, is to reread: "El mismo proceso se produce cuando las personas interactúan con el paisaje" (Astvaldsson, p. 74) ["The process is the same when people interact with the landscape"].

Does the narrative speak for nature? It does and it does not. There is no plea for protectionism or conservation. The jungle was not yet threatened, but it was threatening, mute, and deserted. Initially, I thought hatred would be ever-present in *El entenado*. I also thought *Desaparecer* would have a happy ending. I was wrong, and I was wrong. With each rereading, hatred in *El entenado* is replaced by melancholy, a longing for an impossible return, a longing for a rediscovery of nature. This is never an idealized nature, but the evocation itself, the experience. Like Marlowe in *The Heart of Darkness*, this foundling's destiny becomes to dream this nightmare over and over again. The reason for this is that, besides that nightmare, he has no other story to tell. In *Bubbles*, the German philosopher Peter Sloterdijk writes that "Humans have never lived in a direct relationship with 'nature', and their cultures have certainly never set foot in the realm of what we call the bare facts; their existence has always been exclusively in the breathed, divided, torn-open and restored space" (p. 46). In this case, the restored space is the narrator's pen.

In an interview with Ernesto Zelaya-Miñano, the filmmaker Dorian Fernández-Moris explains how important the jungle was for him in the making of his film:

> Esta historia es por eso muy importante para mí, me interesaba hablar de lo que sé, de lo que he sentido por mucho tiempo por la selva y quería retratarla tal cual. Por mucho tiempo he vivido disgustado por cómo se retrataba la selva de esta forma tan caricaturesca, se iba por el lado exótico principalmente pero no se la retrataba con realidad, con justicia.
>
> [This story is very important for me, I wanted to address what I have felt for a long time for the jungle and I wanted to render it as it is. For a long time I have been disappointed by the caricaturistic rendering of the jungle, its exotic side that failed to show its reality, with justice].

Blaming the mythical fishmen in *Desaparecer* was a ploy. What mattered was the sale and exploitation of rosewood for perfume and cosmetics. Everyone was in on this except the enraged villagers. The lieutenant governor of Nueva Esperanza was working with the capitalists and willing to let the jungle vanish in exchange for money. She is very well aware of this and even stands before an effigy of the Virgin Mary and begs for forgiveness. In this case, the cannibalism, the hate, and the auto-destruction have become trauma. Who survives? Not Milena, not Giovanni, only the nameless narrator of *El entenado*, the foundling with no past, lives to tell. He is the only one who flourishes and emerges from this natural world unscathed, back when the jungle had yet to be exploited. Milena and Giovanni, in turn, lose their lives, a result of being-in-the-jungle. Only the anonymous narrator in *El entenado* had the chance to restore nature and to flourish in what Sloterdijk would call the greenhouse of his autogenous atmosphere (p. 46). Sad tropics, the hate of nature, sadness, nature, and hate coinciding in one title. A line from a poem by Victor Hugo reads, "Flux et reflux. La souffrance et la haine sont soeurs" (Oster, p. 263) ["Flux and reflux, suffering and hate are sisters"].[18]

---

18 Oster, Pierre. *Dictionnaire des citations françaises*. Paris: Robert, 1993.

Image 3: Iquitos, Peru. Houses on the river. https://media.npr.org/assets/img/2014/08/07/dr_peru_place_wide-e5023ec76aa

Bibliography

Abbate, Florencia: *El espesor del presente*. Editorial Universitaria Villa María: Córdoba, Argentina 2014.

Arce, Rafael: *Juan José Saer, La felicidad de la novela*. Universidad Nacional del Litoral: Santa Fe, Argentina 2014.

Astvaldsson, Astvaldur: "Mitos, paisaje, y modernidad en la literatura latinoamericana" In: Chocano / Rowe (eds.): *Huellas del mito prehispánico en la literatura latinoamericana*. Iberoamericana Vervuert: Frankfurt 2011, pp. 67–90.

Corbatta, Jorgelina: *Juan José Saer, Arte poética y práctica literaria*. Corregidor: Buenos Aires 2005.

Fernández, Nancy: *Narraciones viajeras*. Biblos: Buenos Aires 2000.

García Loaeza, Pablo / Garrett, Victoria L. (eds.) *The Improbable Conquest*. The Pennsylvania State University Press: University Park 2015.

Kramer, Nicholas Michael: *Writing from the Riverbank*. ProQuest UMI Dissertation Publishing, 2011.

Laurent, Pénélope: *L'oeuvre de Juan José Saer*. L'Harmattan: Paris 2014.

Lévi-Strauss, Claude: *Tristes Tropiques*. Penguin: New York 1973.

López Baralt, Mercedes: "Viaje a la semilla: la reescritura contemporánea de la letras coloniales." In: Chocano / Rowe (eds.): *Huellas del mito prehispánico en la literatura latinoamericana*. Iberoamericana Vervuert: Frankfurt 2011, pp. 23–39.

Mancini, Adriana, et al. (eds.): *Ficciones argentinas*. Grupo Editorial Norma: Buenos Aires 2004.

Mapangou, Dacharly: *La fiction romanesque de la postmodernité et ses labyrinthes*. Presses Académiques Francophones: Paris 2013.

Oster, Pierre: *Dictionnaire des citations françaises*. Robert: Paris 1993.

Perdigón Torres, Andre: *La littérature obstinée*. Peter Lang: New York 2015.

Pimentel, Sebastián: "*Desaparecer*, una película sin fuerza ni personalidad propia". *El Comercio*, 05/18/2015, from https://elcomercio.pe/luces/cine/desaparecer-pelicula-fuerza-personalidad-propia-363719.

Premat, Julio: *La dicha de Saturno*. Beatriz Viterbo: Rosario, Argentina 2002.

Ramos, Luis: "Entrevista con Dorian Fernández-Moris". Lima 02/10/2015, from https://www.cinencuentro.com/entrevista-dorian-fernandez/.

Ravetti, Graciela: "Los mitos guaraníes sobre el canibalismo y su relación con *El entenado*" In: Chocano / Rowe (eds.): *Huellas del mito prehispánico en la literatura latinoamericana*. Iberoamericana Vervuert: Frankfurt 2011, pp. 386–394.

Rivera-Barnes, Beatriz: "Is there such a Thing as too much Water?". In: Stavans, Ilán (ed.): *Chronicle of the Narváez Expedition*. WW Norton and Co: New York 2012, pp. 204–217.

Saer, Juan José; *El entenado*. Lanzallamas; San José, Costa Rica 2016.

—. *The Witness*. Serpent's Tail: London 1990.

Sarlo, Beatriz: *Escritos sobre literatura argentina*. Siglo XXI Editores Argentina: Buenos Aires 2007.

Schwartz, Jorge: "Saer en la Universidad de São Paulo", from https://www.avizora.com/publicaciones/monosavizor/entrevista_saer.htm

Schwartz, Martha: http://www.marthaschwartz.com/academic/writings/i-hate-nature/

Sloterdijk, Peter: *Rage and Time*. Columbia University Press: New York 2012.

Sloterdijk, Peter: *Bubbles*. The MIT Press: Cambridge and London 2011.

Staden, Hans: *Viajes y cautiverio entre los caníbales*. Nova: Buenos Aires 1961.

Stavans, Ilán, (ed.): *Chronicle of the Narváez Expedition*. Norton: New York 2013.

Underhill, James W.: *Ethnolinguistics and Cultural Concepts*. Cambridge University Press: London 2012.

Wood, Robert D.: *The Voyage of the Water Witch*. Labyrinths: Culver City, CA 1985.

Zelaya Miñano, Ernesto: "*Desaparecer*, a Jungle-set Thriller that Raises the Bar". *Screenanarchy* 05/19/2015, from http://screenanarchy.com/2015/05/review-desaparecer-a-jungle-set-thriller-that-raises-the-bar.html.

Gisela Heffes
Exclusive Natures: Latin American Cities in Urban Ecocritical Perspectives*

Abstract: Today, the problem of contamination and destruction as well as the preservation of natural resources concern us all because as the world is growing increasingly smaller, the elements that harm the environment are becoming increasingly more globalized: just a few examples range from the emission of toxic gases to the pollution of aquifers, either by means of the increasingly frequent use of chemicals in the agriculture industry or the disposal of these components in rivers and oceans. However, the debate concerning ecological sustainability takes on a different dimension when read from those cultural constellations that create dialogue, implicitly or explicitly, with diverse means of conceptualizing the planet's future, posing related questions that have to do with the necessary balance and harmony between the use and preservation of natural forces as well as available resources. This chapter is focused on a wide range of texts as well as documentaries and films, works of art, and urban performances that take place in the city. From an urban ecology perspective, I will examine how urban life is represented vis-à-vis the development of urban environments and/or environmental spaces, spanning from the image of the city as an immense biological organism (Ross and Bennett 1999) to broader issues that include both movements concerned with making cities green as well as works that deal with the distribution of wealth, racial segregation, or waste management in Latin America.

Keywords: Urban Nature, City, Latin American Studies, Ecocriticism

Although ecocriticism is—less and less—an emerging discipline, the relationship between a specific cultural production and the representations of the environment in the space of the city has been even less explored in the last decades. Ecocritics have mostly focused their analysis on spaces or genres associated with what is natural or regional. Two theoretical examples that address the frequent confrontation between the urban and rural are the already classic texts *The Machine in the Garden: Technology and the Pastoral Ideal in American Culture* (1964) by Leo Marx in the United States, and *The Country and the City* (1973) by

---

* This chapter is based on the fourth chapter of my book *Políticas de la destrucción / Poéticas de la conservación. Apuntes para una lectura (eco)crítica del medio ambiente en América Latina*. Beatriz Viterbo: Rosario 2013.

Raymond Williams in Great Britain.[1] Both texts deal with a reading of cultural history that registers the complex relationship between nature and urbanism. While Williams incorporates industrial technology, Marx emphasizes nostalgia for a rural landscape, one that is in an irreversible process of transformation occasioned by economic powers and class interests. In *The Environmental Imagination: Thoreau, Nature Writing, and the Formation of American Culture*, Lawrence Buell points out that Marx and Williams share a cultural and Marxist commitment in which they conceive of the process of modernization as a great— and ironic—narrative of the inevitable triumph of industrial capitalism over and above the countercultures of pastoral opposition, in Marx's case, and local life traditions for Williams (p. 14).[2] For Buell, this indignation from the part of the rural perspective facing the degradation of the landscape has been key to the creation of a "toxic discourse" belonging to the most recent defenders of the environmental justice movement.[3] The emergence of this movement—and its increasing interest and insertion within ecocriticism—coincides with a second phase of ecocriticism as it pertains to promoting a revisionism that reconsiders the organicist models related to the conceptualization of the environment as much as environmental studies, including not only "natural" environments but also those that are "constructed". Both landscapes are historically interwoven and, as Michael Bennett suggests, literary and environmental studies must develop a social ecocriticism that considers deteriorated urban spaces and landscapes as seriously as the "natural" landscape.[4] For an adequate critical practice, it is necessary to integrate the very demands and protests of environmental justice into these representations of "constructed" spaces—and into their direct relationship to issues of preservation and environmental destruction.[5]

---

1 Marx, Leo: *The Machine in the Garden: Technology and the Pastoral Ideal in America*. Oxford University Press: New York 1964; Williams, Raymond: *The Country and the City*. Oxford University Press: New York 1973.
2 Buell, Lawrence: *The Environmental Imagination: Thoreau, Nature Writing, and the Formation of American Culture*. Harvard University Press: Cambridge 2005.
3 See Buell, Lawrence: "Toxic Discourse". *Critical Inquiry* 24(3), Spring 1998, pp. 639–665.
4 Bennett, Michael: "The Social Claim on Urban Ecology" (Interview to Andrew Ross). In: Bennett, Michael / David, W. Teague (eds.): *The Nature of Cities. Ecocriticism and Urban Environments*. University of Arizona Press: Tucson 1999.
5 In *Políticas de la destrucción*, I refer to the need of creating a new critical episteme that take into account—besides a more matured and reflexive environmental ethic and policy—, the intersections of the metropolis and the interior as well as the combination of anthropocentric, biocentric, and ecocentric preoccupations. Heffes, Gisela: *Políticas*

Exclusive Natures   201

In Latin America, urban ecocriticism, or what Bennett refers to as "green cultural criticism", is scattered within a rhetoric of waste (objects and subjects) that needs to be addressed through the works of Michel Foucault, Giorgio Agamben, and Zygmunt Bauman (2004; 2011), among others.[6] These works can turn out to be illuminating for prompting a deeper reflection that addresses literary and cultural phenomena that is conceptually intertwined with what is thrown out, collected, recycled, and preserved in the urban environment, or what I call a biopolitics of waste: in other words, to reflect upon issues such as environmental conflicts emerging from the unequal distribution of natural resources, a consideration of the increasing economic scale regarding the production of waste, the lack of access to sanitary goods and services, and the disproportionate amount of pollution in marginalized populated sectors.[7]

Another question that needs to be considered is that of environmental preservation through the evocation of the utopian imagination. This chapter aims to analyze a corpus of texts and films at the intersection of both a utopian and urban imaginary that belongs to a specific historical phase from the end of twentieth century to the beginning of the twenty-first in Latin America. In the visual and textual narratives I examine here, "natural space" is no longer accessible and available to the entire citizenry; on the contrary, a fragmented space appears that separates and divides its citizens, leaving them in violent confrontational positions on both sides. I have divided this phase into two sections that conflict with each other in an evident—and ironic—way: the first analyzes a corpus of contemporary utopias in which the representation of nature as well as a continuous allusion to certain environmental concerns is explicit and the natural space is visible and exuberant, although restricted to a select minority.[8] These closed

---

de la destrucción / Poéticas de la conservación. Apuntes para una lectura (eco)crítica del medio ambiente en América Latina. Rosario: Beatriz Viterbo 2013.

6   Foucault, Michel: *Histoire de la sexualité*. Vols. 1, 2 & 3. Gallimard: Paris 1976, 1984; Agamben, Giorgio: *Sovereign Power and Bare Life*. Stanford University Press: Stanford 1998; Bauman, Zygmunt: *Collateral Damage: Social Inequalities in a Global Age*. Polity; Cambridge 2011; Bauman, Zygmunt: *Wasted Lives: Modernity and its Outcasts*. Polity: Oxford 2004.

7   These are but a handful of the scholars I draw from in my efforts to configure what I called a "new epistemological archive" (Heffes 2013).

8   By environmental concerns I allude to Carson's references to the "interaction between living things and their surroundings", as well as how "the physical form and the habits of the earth's vegetation and its animal life have been molded by the environment" (p. 4); in Carson, Rachel: *Silent Spring*. Mariner Books: Boston / New York 2002.. However, there are no references whatsoever to "environmental problems" such as "the multiple

utopias, represented in the form of fortified enclaves (private neighborhoods, gated cities), leave a vast number of marginalized people at their exterior. I will thus examine narratives such as the short story "No Retiro da Figueira" (1984) ["At the Fig Tree Retreat"] by Brazilian Moacyr Scliar, the novel *Las viudas de los jueves* [*Thursday Night Widows*] by Argentine Claudia Piñeiro, and the film *La zona* [*The Zone*] by Mexican Rodrigo Plá.[9] These representations forced me, therefore, to read the reverse side of these spaces and analyze the representation of open utopias, or the "outskirts" of these closed spaces, which are nothing other than the present-day Latin American city. I will then conclude this chapter with a brief exploration of Argentine Ana María Shua's novel, *La muerte como efecto secundario* [*Death as a Side Effect*].[10]

The utopian tradition as a literary effort—although also communal and social—declines at the end of the twentieth century. In research about urban utopian representations in Latin American literature (Heffes 2008), I point out that while there are a considerable number of urban utopias that emerge at the beginning of the Latin American modernization process, a symptomatic absence of those alternative formulations characterizes the spatial and imaginary representations that appear in the period after this process.[11] This decline coincides with the decline of the grand narratives and the advent of a neoliberal socioeconomic model. Therefore, I suggest that urban and literary utopian imaginaries constitute a category of analysis in which utopian cities make up a chapter or episode that accompanies the modernization process, and that, when this process comes to an end, these fictional essays likewise disappear. From this conclusion, it is possible to formulate the following question: Does the urban utopian paradigm so characteristic of Latin American modernity reemerge at some later moment after the end of this process, or does this alternative model disappear—that is to say, does it die out of this cultural imaginary forever?

---

forms of ecodegradation that afflict planet" (Buell / Heise et al. 2011, p. 418), in Buell / Heise / Thornber *Literature and Environment. Annual Review of Environment and Resources* 36, 2011, pp. 417–440.

9 Scliar, Moacyr: "No Retiro da Figueira". In: Godoy Ladeira, Julieta de (org.). *Contos brasileiros contemporâneos*. Editora Moderna: São Paulo 2001, pp. 47–50. Piñeiro, Claudia: *Las viudas de los jueves*. Clarín-Alfaguara: Buenos Aires 2005. Plá, Rodrigo (dir.): *La zona*. Prod. Morena Films and Buenaventura Producciones, 2007.

10 Shua, Ana María: *La muerte como efecto secundario*. Editorial Sudamericana: Buenos Aires 1997. All translations from Spanish texts belong to Grady C.Wray.

11 Heffes, Gisela: *Las ciudades imaginarias en la literatura latinoamericana*. Beatriz Viterbo Editora: Rosario 2008.

Moreover, do utopian formulations reappear in which, besides proposing a different archetype, they offer a spatial and social configuration also aligned with environmental premises? On the one hand, the answer is affirmative. On the other, the answer can only be considered through an adversative conjunction: the emergence of urban utopian proposals in which the preservation of nature also forms part of an alleged "green" agenda indeed resurfaces, although through a completely different modality. In the era of neoliberal urbanization, discursive proposals situate their environmental imaginaries in an alternative urban effort made up of real estate projects and undertakings like those found in fortified enclaves, according to Teresa Caldeira's definition that refers to the emergence of private neighborhoods.[12] Even promotional and advertising narratives describe these spaces in many different ways, from urban farms to private neighborhoods, urban condominiums, residential communities, private city-towns, or megaprojects and country club communities (Rojas, p. 15).[13] Additionally, the relationship between urban territory and the environmental agenda, when translated literally in contemporary fictions, reemerges under the configuration of a different spatiality, generally dystopian, in which the idea of ecocide constitutes the epicenter of the stories. Thus, this chapter will concentrate on two contrasting aspects within current representations that use the trope of environmental preservation as a point of reference. On the one hand, the utopian, urban, and environmental discourse, when it reemerges, transforms into a neoliberal utopian discourse that sells nature (and green cities) through the proposal of gated neighborhoods, private country clubs, and city-towns;[14] on the

---

12 Caldeira, Teresa Pires do Rio: *City of Walls: Crime, Segregation, and Citizenship in São Paulo*. University of California Press: Berkeley 2000.
13 Rojas, Patricia: *Mundo privado: historias de vida en countries, barrios y ciudades cerradas*. Planeta: Buenos Aires 2000. This phenomenon is global, and similar paradigms are known as "gated communities" in the United States, "ensembles résidentiels sécurisés" in France, "alcabalas residenciales urbanas" in Venezuela, "fraccionamientos cerrados" around Mexico City, "condominios fechados" in Brazil, and "urbanizaciones cerradas o privadas" on the outskirts of Santiago de Chile (Rojas 2007, pp. 15–16).
14 In a chapter on "green utopias", I compare two utopias of early twentieth century with the intent of demonstrating that urban planning and environmental concerns have intersected the modern Latin American imaginary, even proposing plausible alternatives to detractors of the city, who equated the latter with the core of all the vices derived from the growing industrialization, especially in Europe, thus combining a synthesis between two different ways of conceiving social, metropolitan and—overall—environmental improvement (Heffes 2013: 255). The texts that were analyzed for this purpose are *A través del porvenir. La estrella del sur* (Enrique Vera y González, 1904),

other hand, and outside of this socially self-segregated space, appears another type of representation, in which "green" has disappeared and the perspective that prevails is a dystopian ecotopia or a futurist ecocide. These representations also take place in the urban space and their distinctive traits reconfigure the notion of what has been called "ecological citizenship" (Dobson).[15]

The American historian Matthew H. Edney has suggested that maps constitute complex texts through which human beings organize and communicate their knowledge and their relationship with the world (p. xv).[16] If we carefully examine satellite maps and blueprints designed by architects in charge of real estate projects, private cities, or closed neighborhoods, we can confirm that a shift in the spatial imaginary has been produced that likens these closed communities to those urban models designed and proposed at the beginning of the twentieth century.

By the same token, these private enclaves are rhetorically promoted as green paradises, where natural and social space fuse together and give space to a new universe that did not exist previously and that, in addition, shields the residents from the presumed and growing violence, marginality, insecurity, criminality, and pollution that contaminate and inundate exterior space, or the "outside".[17] The environmental preservation image is evoked in these textual and visual

---

and *La ciudad anarquista americana* (Pierre Quiroule [pseudonym of Joaquín Alejo Falçonnet], 1914).

15 Dobson, Andrew: *Citizenship and the Environment*. Oxford University Press: New York 2003.

16 Edney, Matthew H.: "Cartography Without 'Progress': Reinterpreting the Nature and Historical Development of Map Making". *Cartographica. The International Journal for Geographic Information and Geovisualization* 30 (2), 2011, pp. 305–329.

17 I use the term "presumed" because, although violence has increased in Latin America in recent decades, many attribute this growing perception of insecurity to the media, whose purposes, whether economic or political, are to accentuate it. In fact, the relationship between the "perception of fear and insecurity" and the "need for social distinction can be found in the so-called closed communities or gated communities, as they are called in the United States," whose special mechanisms of "discursive creation" combine the "problem of urban insecurity, the real estate market as city generator, the inability of local governments to organize and provide services and infrastructure, the process of privatization and individualization of public space, the strengthening of so-called consumer society, the weakening of the sense of community, the increase of social and economic inequality, and the symbolic desire to acknowledge social exclusivity on the part of the middle and upper classes, among other factors" (Enríquez Acosta, p. 2). In Enríquez Acosta, Jesús Ángel. *Entre el miedo y la distinción: los fraccionamientos cerrados en la frontera Noroeste de México*. Universidad de Sonora: Hermosillo 2010.

Exclusive Natures 205

representations through a reconfiguration of the spatial imaginary—endemic to the economic implementations of the neoliberal model. This reconfiguration not only fragments the urban social fabric but confronts the "inclusive" with the "exclusive", and intersects with this opposition through a "green" rhetoric in which different versions of human and non-human are likewise pitted against each other.

In order to recount this shift in the configuration of spatial imaginaries and to analyze how an urban and environmental rhetoric can be appropriate for a discourse of profit, I will examine Scliar's, Piñeiro's, and Plá's fictions. All these textual and visual narratives position themselves "within" private utopian formulations and expose the most significant traits of what has become one of the most exclusive paradises in Latin America: the *privatopias*, as was accurately defined by David Harvey.[18] Towards the end of the chapter, I will contrast private neoliberal and contemporary "utopias" with open and unrestricted cities that coexist with the former, and where dystopia and ecocide constitute two of the most distinctive traits. For this purpose I will refer to Shua's novel.

Scliar's short story opens with a description of the place where the main character and his family are going to move. According to the publicity brochure:

> The place was [...] marvelous. Just as the brochure said: marvelous. Full of trees, tranquil, one of the last places—so the ad said—where you could hear the song of a kiskadee. It was true. The first time we went there we actually heard a kiskadee. And we saw that the houses were sturdy and beautiful, exactly as the brochure had described them: modern, solid, and beautiful. We saw the lawns, the parks, the ponies, the little lake. We saw the airfield. We saw the majestic fig tree, the *figueira*, for which the condominium was named: Retiro de Figueira. (p. 48; emphasis added)

This was the gated community whose lifestyle promised two characteristic elements of all these urban projects: nature and isolation. These elements are fundamental since they do not let what is outside infiltrate into the guaranteed

---

18 "The geographical disparities in wealth and power increase to fashion a metropolitan world of chronically uneven geographical development. For a while the inner suburbs drained wealth from the central city but now they, too, have 'problems' though it is there, if anywhere, where most new jobs are created. So the wealth moves, either further out to the ex-urbs that explicitly exclude the poor, the underprivileged, and the marginalized, or it encloses itself behind high walls, in suburban 'privatopias' and urban 'gated communities.' [...] The rich form ghettoes of affluence (their 'bourgeois utopias') and undermine concepts of citizenship, social belonging, and mutual support" (Harvey 2000, pp. 148–150; my emphasis). In Harvey, Davis: *Spaces of Hope*. University of California Press: Berkeley 2000.

security of the residents.[19] Scliar's story recounts the fascination of the main character's wife with all the typical elements of sophisticated security whose protection makes, hypothetically, its inhabitants as well as their belongings inaccessible: from the electric fence to the watchtowers, the spotlights, the alarm system, and, above all, the private security guards, who are always smiling and very nice. The narrative emphasizes the sense of social and economic security for the people who decide to move to these gated communities: the publicity brochures, for example, are only sent to a limited number of people, and the main character is the only one at his workplace to receive one. It is this condition of distinction and exclusivity to which his wife attributes a "careful selection of future residents" that adds "another cause for satisfaction" to the proposal contained in this new lifestyle (Scliar, p. 49). However, Scliar's short story entails a fundamental paradox: the fortified enclosure in which they live turns into a trap since the very same security guards, who are "always smiling and nice", use the residents of the closed neighborhood as hostages to charge ransom that will allow them, as the character says at the end, to perpetuate this real estate system of kidnapping and ransom:

> We never saw the boss or his men again. But I'm sure they are enjoying the ransom we paid. A sufficient quantity to build ten condominiums just like ours—which, by the way, I always thought was very good. (Scliar, p. 50)

On one level, the story's central paradox resides in the fact that the danger that dominates the media discourse is found outside, but it is also found inside, where the characters cannot leave, and it comes back to them in a boomerang effect. As happens in other recent texts and visual representations, life in these green and protected spaces is presented as an attainable utopia, a dream come true:

> We moved. Life there was really a dream. The kiskadees were punctual: at 7:00 in the morning they started their concert. The ponies were gentle, the scrubbed tree-lined avenues were always clean. The breeze blew through the trees of the park—112 of them, just as the brochure said. Also, the alarm system was impeccable. The guards showed up periodically at our house to see if everything was alright—always smiling and nice. Their boss was a particularly dedicated person: he organized parties and tournaments;

---

19 The promise of security is central in these closed territories. In fact, they "base their construction on an image of security in the walls that surround them, the access booths framed by monumental archways and the presence of private guards," and they attribute their success to the "possibility of protecting their residents from insecurity and delinquency that is prevalent in cities" and to the "security measures" that are varied, "but basically [they] rest on the visibility of the elements of surveillance that serve to spark fear or discourage delinquents" (Enríquez Acosta 2010, p. 178)..

he worried about our well-being. He made a list of the relatives and friends of the residents—in case of any emergency, he explained with a calming smile. The first month went by—as the brochure promised—in a climate of dreams. Dreams, indeed. (Scliar, p. 49)

The paradox that this story entails, which establishes a disturbing relationship among utopia, green and natural space, and security vs. criminality, also appears in Plá's film. In fact, the opening image in *La zona* is a wire fence with electrified cables on the upper part, walls with cameras that carefully register the streets and houses, inside and out, and private security guards posted at the entrance gates in control booths paying close attention to the images that multiply on the panels full of monitors. The inside of the private city, with its neighborhoods, schools, and communal spaces for all its residents, appears as an ideal universe, an exclusive paradise with majestic houses, where flora and fauna live together in harmony with urban design and planning. The space is an exclusive and inclusive territory at the same time. On one hand, it excludes the outside, but on the other, on the inside, everyone enjoys the same rights. However, the behavior of its inhabitants presupposes a level of civility that will be questioned at the end of the film. The supposed understanding that socially unifies the residents of this green paradise is broken when the outside permeates the inside: when part of the wall that surrounds the fortification breaks due to the collapse of a publicity tower. Four young people from outside take advantage of the opportunity to penetrate the enclave, or *la zona*. The residents immediately kill three of them in the name of their right to self-defense while the other young person remains hidden, causing a hunt for him. The conflict surfaces when the police from outside want to investigate the shootings, and the private security—with the support of the residents of *la zona*—stops them because of a problem with jurisdiction and the limits of power. Also, there is a typical tendency of the residents of these closed neighborhoods and private cities to avoid publicity at any cost, so as to also solve the problems of criminality and delinquency "inside the gates" (Rojas, p. 79).[20]

The initial understanding that united all the residents of *la zona* begins to break down even more when the board of directors suspects that one of their own residents has violated the implicit pact and contacted the police. Then, a patrol and authoritarian system is set up in which the neighbors begin to watch over their peers. The breakdown that began with the intrusion of the outside

---

20 Inside these enclosed spaces, "there are certain internal and social security devices that allow for the unpleasant to be hidden: not seen" (Rojas 2007, 79).

not only erodes the links established among neighbors, but it also fractures the fabric of diverse family structures. Daniel, one of the main characters in the film and a member of the board of directors of *la zona*, is confronted by his wife, who questions the policies that the board implements in the neighborhood. Additionally, the relationship with their son Alejandro begins to deteriorate when Alejandro discovers that his father, along with the other members of the board, stops the political investigation by bribing a police representative. When the hidden sixteen-year-old is finally found, a furious mob beats him to death. He had repeatedly tried to escape, but the insurmountable walls and surveillance of *la zona* kept him from it. The supposed level of civility that this lifestyle entails (which is translated not only materially through the mansions they live in and the cars they drive, but also in the educational level of the residents whose children attend the most prestigious private schools) becomes undermined by the brutal actions of the "citizens". At the end of the film, the residents throw the young man's violently murdered body in the trash—just as they had previously done with his three companions. These are nameless bodies that will appear in the garbage dumps. Discarded bodies that no one claims; anonymous, dispensable. The unconcerned gesture that characterizes a rhetoric of impunity once again inserts all these narratives into a problem that relates to the environment and new urban configurations but that, from a biopolitics of waste, questions the borders between the human and the non-human. As in Scliar's story, the presumed impermeability of the walls, watchtowers, and private security does not guarantee the absence of crime and violence. On the contrary, violence is frequently unleashed in the very interior of the enclave, either by transfer from one zone to the other, by the "security" itself, or by the members of the closed communities. In this sense, violence is clearly found on both sides of the wall and there is no fence, wall, or wire that can detain it since it is entrenched in the very structure of a society whose social fabric has not only been ripped, but whose ideology has unfolded through the installation of a model of urban segregation and self-segregation that does nothing but deepen even more social, economic, and environmental differences and disparity. Beyond the different sides where people may find themselves, the film exposes how deep-rooted corruption that culturally characterizes this social universe filters through every social membrane and breaks down intimate relationships and interactions among individuals, family members, and fellow citizens.

The cover of the novel *Las viudas de los jueves* shares the image of the wire fence that we saw at the beginning of *La zona*, and it synthesizes life behind the wall, as one of the characters suggests. The novel tells the story of people who live in the closed community Cascade Heights, an exclusive neighborhood marked by middle to upper class social values, although it is almost ready to collapse:

> Cascade Heights is the neighborhood where we live. All us lot [...] Our neighborhood is a gated community, ringed by a perimeter fence that is concealed behind different kinds of shrubbery. It's called The Cascade Heights Country Club. Most of us shorten the name to "The Cascade" and a few people call it "The Heights." It has a golf course, tennis courts, swimming pool and two club houses. And private security. Fifteen security guards working shifts during the day, and twenty-two at night. That's more than five hundred acres of land, accessible only to us or to people authorized by one of us. (Piñeiro, p. 21)

The story revolves around the lives of Virginia, a resident in charge of a real estate agency in the private community of Cascade Heights, her husband Ronie, unemployed for six years, her son Juani, el Tano and his wife, the Uroviches, and other inhabitants of this urban project. All of these characters not only fully identify with a lifestyle that naturally accepts living in an exclusive enclave, but they flaunt it extensively since, as the sociologist Maristella Svampa warns, living this way "full time becomes an invaluable source of social capital" (p. 143).[21] Crime here is domestic, and it occurs at the end of the novel when three of the husbands, who get together every Thursday to drink, play cards, and discuss politics and the economy, commit a murder-suicide staged to look like an accident so their families can claim a considerable life insurance settlement. The novel, which takes place in Argentina at the paradigmatic moment of 2001, allows for a close look at the social framework that characterizes the lives of these people who are deceptive. Keeping up the appearance of a certain standard of living begins to break down. The idea of private neighborhoods as exclusive utopias presents itself, and the idea of "nature" constitutes one of these neighborhoods' most distinctive and, in turn, marketable traits, for which "potential buyers sometimes imagine they have landed in paradise" (Piñeiro, p. 55). The "inside" therefore consists of an exclusive paradise, a walled utopia: "Three hundred houses, with three hundred gardens, with three hundred jasmine plants, contained in a five-hundred-acre estate with a perimeter fence and private security" (Piñeiro, p. 25). The houses are separated from one another by "living fences": bushes in other words. The so-called nature that separates and divides two opposing and exclusive worlds, inside and outside, is an artificial nature, but sufficiently convincing to attract investors, future owners, and all types of clients. Virginia, the protagonist, clarifies: "'natural'

---

21 Svampa, Maristella: *Los que ganaron: la vida en los countries y barrios privados*. Editorial Biblos: Buenos Aires 2008.

because it comprises grass, trees and lakes, not 'natural' in the sense of belonging to a landscape that was here before we arrived. This used to be a swamp" (Piñeiro, p. 70). Like other "utopias of reconstruction" (Mumford), a spatial metamorphosis has been attained, and the former wetlands became an unrecognizable territory, a paradise.[22] Technology and urban design have transformed the swamp into an idyllic territory, to the extent that it is now impossible "to imagine that our fairways were once marshes" (Piñeiro, p. 70). The aestheticization of the natural space, disguised as ecological concern, transforms the water of the river into a "more turquoise green" color, thanks to a water treatment, "and the introduction of certain algae which keep the ecosystem aerated" (pp. 70–71). Nonetheless, to achieve this goal, the fish that were there before the purification have died: "They were undistinguished fish, a sort of bream, brownish-colored. We put in orange perch, which reproduced and became the new masters of the stream." (p. 71)

This procedure ensures the preservation of the flora and fauna considered "apt" for living, resulting in a *superior species* whose logic preserves animals and vegetation that are worthy of existence—although they are judged only by appearance—allowing, as a consequence, "undistinguished" fish to die, who lacked the privilege of existence. The "orange perch" completely take over their habitat, a practice that, it should be clarified, is instrumental in the name of a more "aerated" ecosystem.

The same phenomenon occurs later when "the undesirable presence of packs of stray dogs" that "are entering our communal property" generally alarmed the neighborhood (p. 191, p. 195): "these were dogs without owners, raised in the wild, who came into our compound looking for food. Feral dogs. Not like our dogs, our Golden Retrievers, Short-haired Labradors"—that is, "the breeds most often to be seen out for a walk in Cascade Heights, wearing collars and identity discs engraved with a name and telephone number, lest they stray" (pp. 191–192). "Stray" wild dogs compare to the "undistinguished fish, a sort of bream, brownish-colored". Accordingly, the novel emphasizes the contrast between this utopian, harmonic, and artificially preserved universe and the outside, which is synonymous with chaos, barbarism, and "ugliness".

The impermeability of the huge walls is not enough to stop the continuous flow of social and economic problems. As in Scliar's short story or in *La zona*, the outside and the inside mix together. The wire fence that separates luxury and contemplation on one side, but constitutes the food source for an entire family on the other, cannot eliminate the borders that try to make invisible what is visible

---

22   Mumford, Lewis: *The Story of Utopias*. Boni and Liveright: New York 1922.

and tangible. The spatial disposition of these exclusive cities translates into a self-segregation that impedes all contact with the outside world and, therefore, denies the existence of those invisible subjects. Nonetheless, not only do they live all around these fortified neighborhoods, but also, they enter them on the condition of serving as domestic help. In this role, they dress in uniforms, thereby cleaning up their identities and their presumably dark origins. Thus, the discarded must be considered not only as irrefutable proof of the contradictions of an economic model based on a dynamic of social inclusion/exclusion, but the discarded also constitute a different discursive phenomenon that must be addressed based on the utilization of an alternative apparatus of critical inquiry: a model that incorporates other versions of history—the past, the future, and, above all, the present. What's more, this new paradigm not only integrates disciplines coming from dissimilar fields of study, but also focuses on what is particularly endemic to Latin America, especially the comparison between a socioeconomic—or political—model of total exclusion and new environmental discourses, which have grown increasingly more present. According to the novel's main character, the subjects who work at Cascade Heights live in the "satellite community": a neighborhood with "simple jerrybuilt houses, almost all of them made by the people who live in them—or by their relations or friends" (p. 91). Since the inhabitants of these poor neighborhoods depend on the work that Cascade Heights provides, they are examined daily as they enter and exit across the strictly controlled borders. It even "had been standard to request—confidentially—a criminal record check on gardeners, builders, decorators and any other workers who came regularly to our country club" (p. 83). In effect, as time passes, the residents of Cascade Heights are protected "every more stringently behind bars", and "[p]lans were afoot to replace the perimeter fence with a solid wall, ten feet high:" a wall "so that nobody passing by could look in at us, let alone at our houses or cars. And also so that we did not have to look out." (p. 83) This paradigm of exclusion and social self-segregation not only dilutes the supportive pacts between the subjects who live outside and inside the neighborhood, but, in turn, it erodes the internal ties, since the models in these micro-communities are dramatically reduced and simplified, likewise reducing and simplifying any notion regarding what is outside the walls. Romina, a young woman whose real name is Ramona, tells Juani, Virginia's adolescent son: "Are we shutting ourselves in, or are we shutting out other people so that they can't come in?" (p. 158).[23] As

---

23 The novel succeeds in portraying the values of the residents of the closed neighborhood in a way that I would call laughable: When Ramona is adopted she becomes Romina.

in Scliar's short story and the movie *La zona*, Piñeiro's novel also concludes with a highly militarized confrontation that, in this case, happens with the guards who are "armed with rifles" (p. 274): again the fear that people from the outside will infiltrate and erase the borders that condemn them to live as second class, third class, or classless citizens emerges. Urban fragmentation corresponds to social fragmentation and returns to the subjects who inhabit—but do not share—the same territory and national sovereignty as relentless enemies. As if in an undeclared war, the barriers that divide, segregate, exalt, and minimize the categories and attributes of the inhabitants on both sides, confront them and impede all type of interaction, mobility, intersection, fluidity, and mixing. But the barriers are not so rigid, and the outside infiltrates the inside like the inside infiltrates the outside; although, when this occurs, hate, vengeance, and resentment are so prominent that the outcome is generally traumatic. In sum, these projects symptomatically reconfigure the notion of urban utopia and nature, and this transformation comprises as well the discursive production of the ones who are in charge of designing, planning, imagining, creating, and selling these new urban and green utopias.

In *Mundo Privado. Historias de vida en countries, barrios y ciudades cerradas*, Patricia Rojas compiles a book of chronicles based on interviews conducted in country club neighborhoods, gated neighborhoods and gated cities with the purpose of investigating what life is like inside these fortified enclaves.[24] A paradigmatic case is the Nordelta real estate project in Tigre, located in the province of Buenos Aires. This is an urban archetype of how these closed, idyllic, and private universes are promoted as "natural" spaces and new and ideal worlds, created

---

This change is due to the need, at least in her mother's mind, to erase her origin as well as her dark skin color. In fact, Ramona's mother embodies the prototype of the frivolous middle to upper class woman surrounded by racial prejudice who, facing the adoption situation, cannot resolve her own conflicts and apprehensions with integrity and honesty: the girl's hair "was black, glossy and thick as wire" (p. 40). "The girl might be from Corrientes, but could also be from Misiones, El Chaco or Tucumán. Mariana thought most likely Tucumán. She could imagine that in a few years she would be as sturdy and strapping as the Tucumana woman who cleaned her friend Sara's house" (p. 42). For the mother, the fact that her child cannot be genetically modified is frustrating and, "no matter how much you made her diet or killed her with exercise regimes, she was always going to have thick ankles and Mariana knew that there was no solution for that" (p. 42).

24 Rojas, Patricia: *Mundo privado: historias de vida en countries, barrios y ciudades cerradas*. Planeta; Buenos Aires 2007.

Exclusive Natures 213

and imagined as green urban utopias. The role of nature is fundamental in all of these projects, characterized by "gorgeous views" and "native vegetation" (p. 22). Nature invades the urban space, but it is a pleasant, ordered, rationalized invasion. While it is presented as a space that emphasizes "the defense and consolidation of a way of life" that is "more related to *respecting ecology*, tranquility, and the solace of its members than to simple land speculation", there are no references, as we read in the fictions, to any undertakings on behalf of the betterment of the environment (p. 23; my emphasis).[25] Curiously, beyond the innumerable allusions and references made to terms like "ecology", "nature", "natural", and "green", these evocations function as signifiers that do not directly correspond to the typical elements of environmental discourse. On the contrary, this marketing rhetoric's most immediate aim serves to attract new clients and potential property owners who, besides subscribing to this particular "lifestyle", promote it through the creation of new spatial and social networks.

According to Rojas, what most impressed her when she arrived in Nordelta was that it was in another world: "*a new world without the uncertainties of the previous one*" since there are "no stoplights", "no poor people", "no sidewalks", "no bars on the windows", and the gardens "have exotic flowers", there are "no loose dogs running around", "no clothes hanging on the lines either", and "no trash on

---

25 It's helpful to compare the discourse related to these spaces with Quiroule's green, organic city's project (see Quiroule, Pierre (pseudonym of Joaquín Alejo Falçonnet): *La ciudad anarquista americana*. La Protesta: Buenos Aires 1914). More than one hundred years ago, he proposed a model of urban sustainable life where members of the "commune" were both vegetarian (p. 87) and "farmers" (p. 63). The implied purpose was to establish a close relationship between its inhabitants and nature. In addition, in Quiroule's green utopia, there was an attempt to reduce the "fabrication of paper" and "ink from the presses" (p. 65); the "houses were made of glass and no wood was used in their construction" (p. 66); and the houses had "double walls", filling the empty space between them with "heat resistant substances" (p. 75). According to Quiroule, this architectural design is advanced in that it efficiently preserves energy and avoids unnecessary waste. Not only is electricity used instead of coal, but the inhabited city is also "green", insofar as it forms an "immense park around the industrial city" (which functions as a "lung"); the streets are pedestrian and are surrounded by gardens; the city is clean and healthy, and the air is pure oxygen and not "a horrible compound of vapors and putrefaction" (p. 75). Water was chemically purified and later used to water the crops (p. 74). The energy sources necessary for this society had been sought out in those "natural elements in perpetual motion: wind, rivers, waterfalls, solar heat, etc." (p. 260). A pioneer, Quiroule proposed different energy models: from wind to solar and hydroelectric.

the curbs" (p. 23; my emphasis).[26] The result of this "new world" is that, apart from "everything being new", there are also "developers, ads in the newspapers, and clearly outlined imaginaries about what it must be like to live such a secure, green, happy life", and, in some cases, even the very idea of nature was normalized (p. 26). These ads that refer to nature serve as a discursive strategy to attract new residents (clients) to the megacity. It is always surprising that the entrepreneurs and investors are the ones who determine how the inhabitants of these private cities should live, as well as how nature is regulated—and therefore objectified. The subjects also become objects of manipulation and regulation on the part of the investors, whose purpose is purely lucrative. The ideas laid out about feeding and protecting the animals of the country club are not consistent with any environmental effort or a sincere concern for the preservation of the environment, such as those promoted by foundations and non-governmental organizations (NGOs) that have the protection and conservation of the biotic community and the ecosystem as their principal objective. Rather, the promotion of these neighborhoods is the result of a practice that seeks to enhance the distinctive elements that continually attract new consumers and that, in commercial ventures, transforms nature into an ornamental element, a surplus or an added and differentiating value.

A striking perspective is that of the person in charge of the Pacheco Golf private project, according to whom the "media believes that only members of the elite live in country club communities. And it's not at all like that". He proposes to change this perspective and consider this phenomenon from a different point of view:

> Why don't we look at it from a different angle? Why don't we think about it as an open piece of land where they could've put a slum, some toxic chemical discharge, or a dump. Instead, they've turned it into a place where people can live peacefully and where jobs are generated. Here, there are all types of jobs: from security, gardening, domestic help,

---

26 Inaugurated in 2000, Berti and del Río describe Nordelta as a group of twenty private neighborhoods with a common infrastructure; it covers a surface of 1,600 hectares (3,954 acres) and predicts a population of 90,000 inhabitants by 2020. The megaproject contains education centers at all levels, spaces for recreation and sports, commercial areas, offices, and its own medical and cultural centers. Located thirty kilometers (nineteen miles) from the center of the port city, the project includes a toll way, two train stations, and a heliport, and its "placement on the edge of the Paraná delta allows for planning a navigable connection with Puerto Madero" (pp. 101–102). See Berti, Natalia y Del Río, Juan Pablo. "Disposiciones espaciales y valorización del capital. El caso de Nordelta", *Geograficando*, Año 1, No 1 (2005), pp. 89–109.

architects and construction workers, pool maintenance workers, caddies... Many people don't know that the 200 families who live here generate jobs for 700 people in the zone [...] Tell me: How many factories generate jobs for 700 people without causing contamination and, instead, create local green spaces for sports and living? (Rojas, pp. 144–145)

From the perspective of investors and real estate entrepreneurs, "a slum" is the same as "some toxic chemical discharge, or a dump". The subjects who live in a slum are tangentially dehumanized and objectified, and they acquire the same contaminating traits as chemical products or trash. Real estate entrepreneurs, on the other hand, present themselves as altruistic and selfless in that they not only rescue green space—nature—from human and environmental contamination, but they also promote economic (and environmental) development through the creation of jobs that do not contaminate and that generate "green spaces for the zone". This commercialization of nature disguised as beneficial investment for the economy and the environment, apart from being illusory, is dangerous because it legitimates a false environmental discursivity that labels the inhabitants of informal and precarious human settlements as contaminating elements. Therefore, their eradication or elimination becomes legitimate.[27]

The continuous appeal to nature constitutes one of the most distinctive traits of these discourses that stimulate and promote the consumption of exclusive paradises. Insularity plays a relevant role here, for islands are related to paradisiacal territories. The island, from More's utopian proposal onward, constitutes an archetype of geographical fiction. With the threat located outside these isolated territories, most utopian conceptualizations also advocate for self-sufficiency and oppose commerce and economic interdependence, since the origins of the evils in society are found in these relationships.

Achrony is another distinctive characteristic of utopias: one must add the absence of temporal factors to the absence of a historical dimension that characterizes utopian conceptualizations. Utopia happens in "another place", and oftentimes how one arrives there, the previous social organization, or how the transition to utopia took place is unknown. This trait appears in *Las viudas de los jueves* when Virginia thinks back to the first time she arrived, in a sort of origin myth:

---

27 From an "environmental" perspective, María Carman analyzes two cases in which subjects belonging to the lowest classes and living in informal settlements are labeled as a threat to the environment and equal to toxic and/or contaminating materials. See Carman, María: *Las trampas de la naturaleza: medio ambiente y segregación en Buenos Aires*. FCE; CLACSO: Buenos Aires 2011.

> Those of us who move to Cascade Heights say that we have come in search of "green," a healthy life, sports and security [...] Entrance into The Cascade induces a certain magical forgetfulness of all that went before. The past is reduced to last week, last month, last year [...] Gradually we forget our lifelong friends, the places we once loved, certain relations, memories, mistakes. It's as though it were possible, in mid-life, to tear the pages out of your diary and begin to write something new. (Piñeiro, p. 26)

The present of a utopia is a definitive present that does not change and whose past is unknown; therefore, it is practically nonexistent. Neither is there a future since there is no possible evolution. Utopias are concerned with an eternal present, time that rules perpetually once it is established. In turn, it is an ahistorical time, and generally it involves textual as well as visual representations of a permanent Edenic condition, beyond the laws of historical evolution.

Urban planning constitutes another aspect endemic to utopian conceptualizations: the allusion to the Ideal City recurs in utopian thought, which proposes the Ideal City as an alternative to the real city since in real cities all evils and threats are condensed and these evils are precisely what should be evaded. Nordelta—as well as the vast majority of private neighborhoods, country club communities, and closed cities—is the result of careful planning whose purpose is "a world apart [...] The most complete country club in the country, with *maximum comfort, surrounded by green spaces and groves of trees that have been here for years*. Wide streets with *all the services of permanent security*" (Rojas, p. 292; my emphasis).

Besides security, a high number of regulations also distinguish utopias. Generally, utopia is totalizing in that it looks to organize social harmony through an integral theory in which all the aspects of collective and private life must be planned. Therefore, utopia tends toward collectivism, homogeneity (which is the result of rigid regulation of daily life, work, and free time), and authoritarianism. This regulation, a consequence of the search for something definite and stable and a result of the eternal and ideal present, leads to the formulation of a social structure in which, apparently, problems have been permanently solved. This regulation reigns over public as well as private life, and its guiding principles can often fall into dogmatism and/or totalitarianism.[28]

Given the characteristics of these fortified enclaves, we can suggest that—as an emerging model that has grown in recent years—they shape a new social

---

28 In fact, this condition has provoked the inversion of utopian discourse into the so-called counter-utopias, anti-utopias, or negative utopias that proliferated during the twentieth century beginning with the writings of Eugene Zamiatin (*We*, 1924), Aldous Huxley (*Brave New World*, 1932), and George Orwell (*Nineteen Eighty-Four*, 1949).

paradigm in that they combine two fundamental spatial elements: the urban and the natural. This imagination, conceptualization, projection, and spatial configuration, when it becomes a specific discursive phenomenon, shares more significant and defining traits of utopian formulations, as we have already indicated. However, two significant characteristics distinguish the discursive rhetoric that cuts through the early twenty-first century utopian undertakings: the initial formulation of these proposals considers exclusive spaces where a select minority can live, and this minority constitutes less than ten percent of the national population. To be a "member", one must be "admitted" by a board that interviews the interested families. If they are lucky enough to be accepted, they must pay a higher initial fee—in this way, families whose salaries cannot cover the costs of living in this upper-class space are filtered out.[29]

The present utopian model, which likens urban space to nature, is a model of social exclusion and segregation. It physically and symbolically reinforces the differences between those inside and those outside. These reinforcements provide their urban residents a particular identity as a special, privileged caste, separated from the rest. Referring to the well-known work of Bourdieu (1979), Maristella Svampa (2008) maintains that "country club communities are spaces of production of *'strategies of distinction'* par excellence", in the sense of "social and cultural standards and practices that make up different status groups" (p. 126; my emphasis).[30] However, the crucial question that cuts through these formulations of urban paradigms is to what extent a city can be privatized and, as a consequence, close itself off and retreat into itself. If this happens, not only does urban space become fragmented, but also the very notion of citizenship is broken. In this sense, Svampa refers to "inherited citizenship", which displaces a model of political citizenship, because it is supported by universal criteria, it therefore has a more general scope (p. 204). Based on this transformation of citizenship, it is fitting to wonder if it is possible to construct a true "social pact" on the basis of inherited citizenship (p. 205).

The second characteristic that differentiates and distances these ideal green urban spatial proposals from other utopian formulations is that the

---

29 "Through a code of restrictions", the "conditions for admission, which are generally unwritten but basically known by all" make up "much more than explicit rules," especially the "profiles of those who can belong" (Svampa, pp. 126; original emphasis). See Svampa, Maristella: *Los que ganaron: la vida en los countries y privados*. Editorial Biblos: Buenos Aires 2008.
30 Bourdieu, Pierre: *La distinction: critique sociale du jugement* Éditions de Minuit: Paris 1979.

environmental perspective articulated and promoted at the moment of defining and giving form to the urban imaginary of these closed paradises is artificial. The environmental references turn the green—as well as all associations linked to nature—into a commercial rhetorical object that appeals to supposed environmental paradises with the exclusive purpose of profit. In this way, one can also legitimately question to what extent the appeal of these green models is sustainable, precisely from an environmental perspective. Therefore, I would like to argue that urban enclosures in Latin America are represented as "sustainable" spaces by recurring to a rhetoric that constantly and interchangeably makes use of terms such as green, natural, and nature. Furthermore, in the context of urban closed enclaves, the reference to terms such as "green" or "nature" is used as an empty signifier whose referent (or lack thereof) is not anchored to a real environmental concern and therefore changes according to the possibilities that the market dictates.[31] The term depends on supply and demand, transforming itself into a word that only serves the effects of economic profitability. So how do the concepts of "green" and "nature", so characteristic of the rhetoric inherent in the representations of closed and exclusive spaces, appear when we read them from the other side, the outside, or backwards? What natural elements are visible in the Latin American narratives of the future whose epicenter, the open urban space—in opposition to the closed and fortified city—is available for everyone and produced by everyone? To answer this questions, I will analyze the Argentine writer Ana María Shua's novel *La muerte como efecto secundario* (1997), which I will also compare to the previously analyzed texts.

Shua's story takes place in a transformed Buenos Aires, possible and therefore utopian. But, contrary to the textual and visual representations that we have just analyzed, this Buenos Aires evokes a dissimilar future with respect to any idea of happiness, refuge, tranquility, and safety: traits that constitute the discursive and identifiable markings of discursive phenomena related to private cities. The novel revolves around Ernesto Kollody and his tyrannical father who, old and infirm, must be admitted to a "Convalescent Home". This space is obligatory for all the old people in the city, and additionally, illness and agony are prolonged in this facility exclusively for economic purposes. In this possible Buenos Aires, one

---

31 My argument doesn't claim that just because it is not driven by an environmental concern, it means that it is unintentionally ecological regardless. However, in all these narratives, there is a symptomatic absence of references that address, for instance, the aftermath of populations that have been displaced in order to develop these multibillion projects; the generation of waste and the lack of recyclable initiative; and the problem of land deforestation, erosion, pollution and contamination.

cannot walk in the city, only in "shopping centers or in guarded neighborhoods", and there are "many walking tracks in the city" and "secured places that pretend to be ordinary neighborhoods where, for a modest fee, it's possible to wear yourself out walking, passing infinite—or finite—landscapes, almost real" (p. 7).

Open space, or what is outside of the secure, natural, protected, and surveilled inside, consists of ruined, deteriorated, and dangerous areas that one can only cross in taxis and bullet-proof vehicles. The fenced-in and sealed-off city contrasts with what is outside, ruins and rubble, and makes the fortified space on the inside into a controlled space, free from any possible unknown incident. There is generalized violence; vandalism and crime have spread through the urban areas, and now the people who carry out such acts represent all ages and genders, as well as a huge variance in levels of professionalism and motivation. Besides the possibility of openly traveling through the city, other things are lost in this dystopian universe, such as cultural production. A language devoid of authenticity must be added to the death of cultural production. The Convalescent Home, for example, is a way to euphemistically refer to what used to be old age homes, geriatric centers, and senior residences. However, not all of the Convalescent Homes were the same. Just as closed neighborhoods flaunted vegetation and nature, prohibited to those who remained on the outside, the Convalescent Homes also adhered to the private/public logic.

In Shua's tale, Ernesto rescues his father from the Convalescent Home by contracting a commando group who lives in another highly demarcated space, although this one—unlike the closed neighborhoods—is not found on any map: these spaces are called "occupied zones". This symbolic and extra-official geography has not only been removed from the most updated urban maps, but it now appears "as if they were parks or plazas that one has to circle around" (p. 112). The natural space that remains for the wealthy is here transformed into another empty reference in that its cartographic demarcation does not correspond to its real condition (which is erased). The occupied zones are far from the "parks or plazas", false representations of green spaces that the city lacks or reserves only for those who have the economic means to live among their own. The occupied zones are characterized by "physical deterioration", houses, and buildings that "experience a process of degradation that simple poverty can't explain" (p. 113). This space represents a chaotic cartography that evokes a rhetoric of waste that appears between the folds and interstices on the edges and borders that limit the different zones (of contact and intersection). The most recognizable opposite of these zones is the high-resolution map of the private neighborhoods, the "privatopias", where the spatial configuration is clearly marked off and defined, always with the color green that surrounds all the urban constructions, as in the maps of utopias.[32]

With his father rescued from the Convalescent Home, Ernesto embarks with him on an escape that, after cutting through fragmented spaces of the city, ends in the northern zone of the province of Buenos Aires. In this city of the future, even outside the limits of the capital, the space was urbanized and "areas of no-man's-land were practically nonexistent"; on the contrary, one could see "gated communities, occupied zones, villas, and nothing else" (p. 151). Ernesto and his father go to another exclusive space, another green utopia, which is also a mythic space inhabited by "Old Runaways", old people who have escaped from the Convalescent Home and who have founded their own utopian community. In fact, this "community of Old Runaways" or spatial myth invites belief in a gerontopia in a particular moment when the elderly are not only economically exploited, but when the Convalescent Homes cover up social gerontophobia: the need to cloister off the old people under multiple pretexts allows for easily avoiding the problem that old age carries in contemporary society by making the elderly more invisible. Finally, Ernesto and his father find it. The Old Runaways had taken over the prestigious and exclusive Highland country club community, turning it into an occupied zone, a utopia only for themselves.

If, in these contemporary textual and visual representations, the open and available space of the city is a dystopian space, then in Shua's narration, only private neighborhoods conform to urban and green utopias, increasingly resembling the exclusive paradises of private neoliberal utopias. The open dystopian space is marked by wars between private armies that take the place of the completely absent State and by the lack of nature and green space. The reference on the maps to the color green only disguises the continuous degradation to which one is subject in this possible Buenos Aires. Just like another artificiality among all the illusions that abound in this text, green substitutes for what really unfolds in the occupied zones. The city now divides urban space into green and private territories for the few and leaves gray and degraded spaces "democratically" open and available for all, as Ulrich Beck (1992) ironically refers to poverty and pollution.[33] Therefore, these spatial representations must be read in the same

---

32 As in the films *La zona* and *Una semana solos*, and also in Ariel Winograd's *Cara de queso* [*Cheese Face*] (2006), nature appears, characterizes, and defines one of the most important specificities of country club communities. The repeated traveling aerial shots expose a sort of precise map on which each house is surrounded by a lush lawn.
33 Beck, Ulrich: *Risk Society: Towards a New Modernity*.: Sage Publications: London; Newbury Park, CA 1992.

way as the demands of environmental justice, in which environmental disparity combines with social, cultural, and economic inequality. The only exception regarding this urban paradigm is the community of the Old Runaways: a utopia that, due to its mythical condition, only exists ambiguously.

To this contrast, we must add that the outside, besides being permanently exposed to continuous deterioration and attacks, is plagued by persistent climatic erosion sedimentation—especially excessive heat, to which there is constant allusion, although tangentially. Heat, drought, the disappearance of umbrellas, the hole in the ozone layer: all these references throughout the text are an invisible presence, although by no means insignificant. The corrosive effect of the high temperatures paradoxically permeates the characters' lives, and this aspect is absent from the discursivities of closed cities and fortified enclaves that promote natural, green spaces and nature. It seems that private utopias, because of their very condition as paradises, can nullify global problems such as climate change, global warming, pollution, deforestation, desertification, or the problem of endangered species.[34]

In conclusion, these urban utopian proposals are made up of a criticism of reality, but their visions of the future vary, as do the relationship between the search (or lack thereof) for an alternative social model, the implementation of these imaginaries in a natural space, and the uses (or lack thereof) of its resources. In these narratives (textual and visual), the city is the epicenter of these transformations, but while the first ones represent urban, green, and exclusive propositions that are conceptualized—and commercialized—as utopian spaces where the dream of openness and social inclusion remains truncated. In spite of being urban, they are characterized by being relegated to an exclusive minority and by equating nature and environmental concern with aesthetic and profitable aspects. They do not, on the other hand, complement or relate these aspects with a more critical method of environmental inquiry. It is not a nature whose environmental agenda seeks to implement a sustainable urban model that raises consciousness of the possibilities of improving its physical surroundings—in broader terms, the ecosystem. Therefore, the concept of nature functions as an empty signifier whose possible referents vary in accordance with first, the supply and demand of the market; second, the development of real estate projects; third, the exacerbation of problems such as (the lack of) safety portrayed through mass media and alliances of economic and political power among entrepreneurs and

---

34 In fact, the aspect of imminent ecocide relates Shua's text to Homero Aridjis's novel ¿*En quién piensas cuando haces el amor?* (1995).

exploitative public officials; and fourth, the imposition—through mechanisms of social distinction—of a change in the supposed "lifestyle" of consumers (who, besides consuming Coca-Cola, consume nature, among many other things). In Shua's novel, on the contrary, the open territory of the city emerges as an "inclusive" space, dystopian, and lacking nature and green, and at the edge of a structural collapse. This aspect of the representations of external spaces make problematic the viability of the utopian proposals at the beginning of the twenty-first century within a context of marked self-segregation and economic disparity. While the utopian impulse is not dead, utopian urban territories are yet to reemerge with new, original—and more inclusive—proposals.

Bibliography

Agamben, Giorgio: Homo Sacer. *Sovereign Power and Bare Life*. Stanford University Press: Stanford 1998.

Aridjis, Homero: *¿En quién piensas cuando haces el amor?*. Alfaguara: Buenos Aires 1995.

Bauman, Zygmunt: *Collateral Damage: Social Inequalities in a Global Age*. Malden: Cambridge 2011; Polity: Malden, MA 2011.

Bauman, Zygmunt: *Wasted Lives: Modernity and its Outcasts*. Oxford: Polity 2004; Blackwell: Malden, MA 2004.

Beck, Ulrich: *Risk Society: Towards a New Modernity*. Sage Publications: London; Newbury Park, CA 1992.

Bennett, Michael: "The Social Claim on Urban Ecology" (Interview to Andrew Ross). In: Bennett, Michael / Teague, David W. (eds.): *The Nature of Cities. Ecocriticism and Urban Environments*. University of Arizona Press: Tucson, AZ 1999.

Berti, Natalia / Río, Juan Pablo del: "Disposiciones espaciales y valorización del capital. El caso de Nordelta". *Geograficando* 1(1), 2005, pp. 89–109.

Bourdieu, Pierre: *La distinction: critique sociale du jugement*. Éditions de Minuit: Paris 1979.

Buell, Lawrence: *The Environmental Imagination: Thoreau, Nature Writing, and the Formation of American Culture*. Harvard University Press: Cambridge, MA 1995.

Buel, Lawrence: "Toxic Discourse". *Critical Inquiry* 24(3), Spring 1998, pp. 639–665.

Buell, Lawrence / Heise, Ursula K. / Thornber, Karen: *Literature and Environment*. Annual Review of Environment and Resources 36, 2011, pp. 417–440.

Cabrales Baraja, Luis Felipe (ed.): *Latinoamérica: países abiertos, ciudades cerradas*. Universidad de Guadalajara/Unesco: Guadalajara 2002

Caldeira, Teresa Pires do Rio: *City of Walls: Crime, Segregation, and Citizenship in São Paulo*. University of California Press: Berkeley 2000

Carman, María: *Las trampas de la naturaleza: medio ambiente y segregación en Buenos Aires*. FCE; CLACSO: Buenos Aires 2011.

Carson, Rachel: *Silent Spring*. Mariner Books: Boston / New York 2002.

Dobson, Andrew: *Citizenship and the Environment*. Oxford; New York: Oxford University Press, 2003.

Huxley, Aldous: *Brave New World*. Vintage: London 2004.

Edney, Matthew H.: "Cartography Without 'Progress': Reinterpreting the Nature and Historical Development of Map Making". *Cartographica. The International Journal for Geographic Information and Geovisualization* 30(2), 2011, pp. 305–329.

Enríquez Acosta, Jesús Ángel: *Entre el miedo y la distinción: los fraccionamientos cerrados en la frontera Noroeste de México*. Universidad de Sonora: Hermosillo 2010; Cengage Learning: México, D.F. 2010.

Foucault, Michel: *Histoire de la sexualité*. Vols. 1, 2 & 3. Gallimard: Paris 1976, 1984.

Harvey, David: *Espacios de esperanza*. Ediciones Akal: Madrid 2007.

Harvey, David: *Spaces of Hope*. University of California Press: Berkeley 2000.

Harvey, David: "The Right to the City". *New Left Review* 53, September–October 2008, pp. 23–40.

Heffes, Gisela: *Las ciudades imaginarias en la literatura latinoamericana*. Beatriz Viterbo Editora: Rosario 2008.

Heffes, Gisela: *Políticas de la destrucción / Poéticas de la conservación. Apuntes para una lectura (eco)crítica del medio ambiente en América Latina*. Beatriz Viterbo: Rosario 2013.

Marx, Leo: *The Machine in the Garden: Technology and the Pastoral Ideal in America*. Oxford University Press: New York 1964.

Mumford, Lewis: *The Story of Utopias*. Boni and Liveright: New York 1922.

Orwell, George: Penguin: London 2008.

Piñeiro, Claudia: *Las viudas de los jueves*. Clarín-Alfaguara: Buenos Aires 2005.

Plá, Rodrigo (dir.): *La zona*. Prod. Morena Films y Buenaventura Producciones, 2007.

Quiroule, Pierre (pseudonym of Joaquín Alejo Falçonnet): *La ciudad anarquista americana*. La Protesta: Buenos Aires 1914

Rojas, Patricia: *Mundo privado: historias de vida en countries, barrios y ciudades cerradas*. Planeta: Buenos Aires 2007

Scliar, Moacyr: "No Retiro da Figueira". In: Godoy Ladeira, Julieta de (org.): *Contos brasileiros contemporâneos*: Editora Moderna: São Paulo, Brasil, pp. 47–50.

Shua, Ana María: *La muerte como efecto secundario*.: Editorial Sudamericana: Buenos Aires 1997.

Svampa, Maristella: *Los que ganaron: la vida en los countries y barrios privados*. Editorial Biblos: Buenos Aires 2007.

Vera y González, Enrique: *A través del porvenir. La estrella del sur* [1904]. Instituto Histórico de la Ciudad de Buenos Aires:Buenos Aires 2000.

Williams, Raymond: *The Country and the City*. Oxford University Press: New York 1973.

Zamyatin, Yevgeny: *We*. Penguin: London 1993.

Manuel Silva-Ferrer
Petrofictions: Nature and Imaginaries of Oil in Latin America

> The eye in oil's dim gas
> was extinguishing around the houses.
> The eye in the dim gas was dying.
> Your name of boy-Cabimas.
> Bus-Cabimas. Fish-Cabimas
> was returning from a trip into the future
> from another time that kept you in the distance.
>
> *Hesnor Rivera. Cabimas*

Abstract: Petroleum is a topic that has been extensively explored by researchers and scholars from all fields of science; however, what has been written in literature and cultural studies about oil is relatively insufficient. This chapter intends to join those that have sought to fill that void through an approach to the discourses and representations around the so-called black gold in Latin America. It is my aim to analyze how certain literature that was produced throughout the twentieth century reflects the radical sociocultural transformations that occurred after an accelerated process of incorporation into modernity driven by the oil industry. This epochal change marked the beginning of a new phase of globalization, determined by the rise and expansion of fossil energies. In this context, I would like to turn my attention to the "ecological turn" started two decades ago in the fields of literature and humanities. Consequently, I wish to pay attention to the call made by Cheryll Glotfelty (1996) in her introduction to *The Ecocriticism Reader: Landmarks in Literary Ecology* to introduce some environmental considerations into the analysis of the texts. I do not pretend to develop a strictly ecocentric approach, an increasing trend in literature and humanities. The goal is, above all, as proposed by Gisela Heffes, to involve the studies of literature and culture with global ecological issues. This theoretical alternative seeks to link the humanities with the environmental movement in the Latin American context. By focusing on this work in texts produced around the landscapes of Latin American oil, I will try to show the special counterpoints between modernity and environment, as well as the contradictions between local struggles and global flows that reflect such writings, while aiming to relocate a portion of Latin American literature and criticism within the new perspectives that have begun to be explored in the humanities and social sciences.

Keywords: Petrofiction, Nature, Latin American Literature, Ecocriticism

Introduction

Petroleum is one of those topics that has been extensively studied by researchers and scholars from all fields of science; however, it is generally considered that what has been written about oil in the fields of literature and culture is rather lacking. This chapter, part of a much broader study, attempts to show that the discourses and representations regarding the so-called black gold in Latin America are not as immaterial as is commonly believed. It is my aim here to show that certain literature produced throughout the twentieth century reflects the radical socioecological transformations that occurred after the accelerated process of incorporation into modernity driven by the oil industry; an epochal change identified with the beginning of a globalization phase determined by the upsurge and expansion of fossil energies.

In this context, I would like to turn my attention to the "ecological turn" begun two decades ago in the fields of literature and humanities. Accordingly, I wish to call attention to the petition made by Cheryll Glotfelty (1996) in her introduction to *The Ecocriticism Reader: Landmarks in Literary Ecology* to introduce environmental considerations into the analysis of the texts.[1] I do not pretend to develop a strictly ecocentric approach, an increasing trend in literature and humanities. My goal is, above all, as proposed by Gisela Heffes (2014), to engage the studies of literature and culture in global ecological issues; this theoretical alternative seeks to associate the humanities with the environmental movement in the Latin American context.[2]

By focusing on texts produced around the landscapes of Latin American oil, most of them in Venezuela, the Latin American petro-state par excellence, I will try to analyze the particular counterpoints between modernity and environment, as well as the contradictions between local struggles and global currents that reflect such writings; likewise, I will attempt to relocate a portion of Latin American literature and analysis within the new perspectives that have begun to be explored in the humanities and in social sciences. This last aspect is of great importance, given that Latin America has been scarcely visible in present-day international debates regarding the so-called petroculture, or petrofictions,

---

1 Glotfelty, Cheryll / Fromm, Harold: *The Ecocriticism Reader: Landmarks in Literary Ecology*. University of Georgia Press: Athens, Georgia 1996.
2 Heffes, Gisela: "Para una ecocrítica latinoamericana: entre la postulación de un ecocentrismo crítico y la crítica a un antropocentrismo hegemónico". *Revista de Crítica Literaria Latinoamericana* 79, 2014, pp. 3–6.

although the region has one of the longest and richest traditions in the production of representations and imaginaries related to the issue of oil.

The First Big Oil Hunt: Chronicles of a New Discovery

-Oil! shouted a man-
and the bellow went through the town like a spark on a
powder keg.
-Oil!, -they all shouted,-
and the same thrill at the sentinel's shout of land was felt.

During the latter part of his life when Alzheimer's disease forced him to live a life of hardships, my grandfather's muddled memory could hardly recall the time of the pioneer oil workers in the townships that border Lago de Maracaibo. During those first years, Cabimas was a tiny hamlet stuck on a blistering coastline under a pristine maritime sun, consisting of palm shacks strewn haphazardly among mud and brush. The trails and swamps traveled on by a few half-naked peasants fleeing the malaria-infected mosquitoes could hardly be called streets, or even roads.

Affected by an illness that held him for long hours, at times for days on end, absorbed in his own irreality, my grandfather's memory would sporadically surface from the lethargic state in which he had plunged, and he tried to retrace the stages of his life that had taken place on his return to Venezuela in 1912, a time during which the banks of the lake had become a true boom-town, and world pressure in the search for oil was in full swing; it became a sort of holy land for fortune-hunters who arrived drawn by countless tales of fantastic wealth crafted around the oil.

In the jumble of my grandfather's soliloquies, it was difficult to distinguish between his senile fantasies and that which quite possibly formed part of his own experience in Cabimas. He first worked as an assistant to a group of young geologists—almost all of them from Stanford University, under contract to the Venezuelan Oil Concessions—and later as doctor for the Creole Plant. He had learned English as a young man during a sojourn in Trinidad; this ability became a tool almost as important as his degree in Medicine and the one which would enable him to enter into what was known at the time as "the industry": small camps entrenched in the jungle, where he worked not only as a doctor but mostly as a cultural translator in a sort of tropical Babelia to which people from all over the world had rushed after the oil fever exploded among the Pedernales quagmire and the northern Guajira.

That was Cabimas—my grandfather used to say—untilled land of white and blue palmares, teeming with thieves, beggars, prostitutes, and people of strange languages. It was a remote place to which an outlandish American man had come—a very refined gentleman, he used to say—who, in a suit and tie in the midst of that blue inferno at 38 degrees in the shade, had dedicated his life to record in countless notebooks all sorts of data of those lost landscapes, ignored by civilization up until that time. My grandfather couldn't remember how he had met that restless young man with whom he had become friends thanks to the English he had learned in his youth, but each time he wanted to recall his name, which happened often, he would remain for long periods with his gaze fixed on some corner of the house, as if looking in the walls for his memory. All of a sudden, a spark would illuminate his mind, and he would stand up and say, striving to pronounce it correctly: Arnold, his name was Ralph Arnold.

I don't know when the name of the American friend, and those of the other geologists and engineers I had listened to for so long, had been destined to one of those drawers of the past in which old tales and photographs intermingle as matters of family memory. Perhaps it would have remained there forever, if not for the fact that several decades later, set to the task of exploring oil matters, I ran into an extraordinary book called *The First Big Oil Hunt: Venezuela 1911-1916*, published in 1960 by the same Ralph Arnold together with his colleagues George A. Macready and Thomas Barrington.[3]

The First Big Oil Hunt consists of a set of accounts organized chronologically that tell of the most extensive search for oil carried out at the beginning of the twentieth century on the American continent. It is a description that presents the socioeconomic conditions that prevailed throughout the northern part of Venezuela and southern Trinidad, as well as the political and material circumstances of the era during which this work was carried out. It is, essentially, a set of hybrid texts—whose vocabulary is halfway between soil engineering and travel stories—that compile the stories my grandfather used to tell of his experiences with the young American geologists that carried out those first expeditions.

How much reality and fantasy are there in these diaries? How many of these tales have for decades fed the stories and imaginary of the nature constructed around the oil scenarios?

---

3   Arnold, Ralph / Macready, George / Barrington, Thomas: *The First Big Oil Hunt. Venezuela 1911-1916*. Vantage Press: New York. 1960. The book has a second edition published in Spanish in 2008: *Venezuela petrolera: Primeros pasos. 1911-1916*. Fundación Trilobita: Caracas 2008.

As we know, memory not only feeds from our own recollections but also from that of others, reinforcing and amplifying our own imaginary account from which memory is articulated. Hence, paradox, one of the most attractive aspects of *The First Big Oil Hunt*, is established when the text, written from the perspective of a science as anchored in the soil as geology, inevitably develops an imaginary account, a fiction, through its reconstruction of the oil scenarios that begin with the recollections of others. The authors thus warn in their introduction: "In many instances, the authors had to depend almost entirely on their memory; in other cases, notes, reports, photographs, maps and letters to the family build a platform on which the story rises" (1960, p. 19).

In this sense, when searching for references to Arnold, Macready, and Barrington's account, what first comes to mind is the chronicles of the Indies. These texts were first produced in the fifteenth century by monks and conquistadors to describe—in the same manner these modern engineers do—their experiences in the process of encountering nature on what is today the American continent, its conquest, and colonization. Nonetheless, the connection with the works on nature exploration by Alexander von Humboldt and Aime Bonpland is evident in that same area. For this reason the scholars were often described in history and literature as "America's second discoverers".

Then, perhaps the fact that the expeditions of Colón, Humboldt, and Arnold all began on the same eastern coast of Venezuela, in the Golfo de Paria, is not just a simple geographic coincidence, nor is the fact that their protagonists experienced the same breathtaking dazzle when they faced with an exuberant nature a radically new world they were meant to explore, get to know, rule, and take advantage of. In the words of the engineer Aguerrevere:

> Very much like a new discovery—in a way like a mock re-enactment of 1492—I don't know which of the two groups, the Venezuelan or the American boys, were more bewildered in that new meeting of two civilizations, two points of view, two traditions, two systems. (Arnold / Macready / Barrington 1960, p. 326)

This perspective of the other, expressed by the young Aguerrevere, outlines the point of view that prevails in the testimonials of the novel explorers who do not hesitate to consider themselves the discoverers of a new world. And in a certain manner, they were, given the conditions of the territory they confronted: "there were no maps worthy of the name to be found in most of the regions explored. Practically nothing was known […] We had to start from scratch" (Arnold et al., p. 19).

Thus, as if taken from the pages of one of the travel logs, the text recreates the initial days of the geological explorations and topographical survey of a vast region where oil was believed to be found. A project, drawn with geographic precision and a language rich in technicalities, depicts the beginnings of the deforestation work of the South American forests for the utilization of the subsoil for the installation of the first oil wells and oil fields—"what it entailed was the optimization of the investment. It was necessary to produce abundant oil and asphalt at the least cost possible". At the same time, as in Conrad's *The Heart of Darkness*, the account of the demanding explorations in a savage territory that went from the southeast coast of Trinidad to the Sierra de Perijá on the Colombia-Venezuela border, is narrated in the first person. Consequently, the accounts are filled with personal suffering and loss of human lives, the battle with tropical illness, and the dynamics of a world in full effervescence: the oil spills, the frequent accidents and fires, the recuperation of pipelines, and of course, the oil that inundates the forest, rivers and lakes.

In one of his accounts, George Macready mentions the letter sent to him by Alfred Schultz's wife some years after his departure, telling him of the experiences in Guanoco field on the eastern border of Venezuela:

> In January 1911, Guanoco was almost entirely destroyed by fire. It started in the village butcher shop, we were told, and it swept through the village, finally igniting a building used for storing explosives. Several natives were injured in the explosion. The fire left the village in ashes. The foreman at the Guanoco Pitch Lake, a man named Hendrie, was injured too, and later died of his injuries. After only two months, my husband was hospitalized with typhoid malaria, and after three months of treatment there, we went home. (Arnold et al., p. 29)

In this sense, the descriptions in these accounts report that things were changing fast and drastically not only on the surface. Indeed, the detailed report of the geological works reveals the consequences that the exploitation of oil was having in the very depths of the earth. Contamination and deforestation thus encompass the very modification of the interior structure of the planet, described in the texts almost as routine in geological work:

> Using his glasses, Harrigas saw "two torches emanating from an island. [...] One torch was at the northern of the island, the height of the flame being about 300 feet; [...] four days later Macready went by boat to the island and found it was another mud volcano; this one, however, about 8 1/2 acres of mud and rock with traces of oil tar; a total of about 696,960 tons of earth that had been blown out of the lower Tertiary strata through two craters by an enormous explosion of petroleum and gas". (Arnold et al., p. 30)

Together with Arnold, Macready and Barrington's book, there is another, nearly unknown work published and translated into Spanish under the name *Los Antecesores: orígenes y consolidación de una empresa petrolera* [*The*

Predecessors: Origins and Consolidation of an Oil Company]. In this case, it is a document by an anonymous author, without a title, originally written in English and published in 1989 in Spanish by the Venezuelan oil company Lagoven. Covering a longer period of the history of the so-called pioneers, this book must have been written at the same time as *The First Big Oil Hunt* or perhaps, due to its similarities, it could even be a part of it, as *Los antecesores* is also a great narrative that seeks to highlight the work of the industry pioneers. Nevertheless, it was left in the archives of Lagoven for a long period of time during which Arnold was forced to wait to finally publish his work and make known the reports that, in his time, had an inestimable value for oil companies.

In any case, due to their localization, subject, structure, and formal features, both works make a sort of diptych that contains the most valuable narrative reservoir of the expeditions that resulted in the discovery of the first large-scale oil fields on the American continent: "This being the first systematic geological exploration on a nationwide basis" (Arnold et al., p. 57). This adventure unexpectedly turned the area into an essential node of a new global ecology determined by oil, and of the vertiginous bioenergetic and socioecological transformation that occurred as of that moment.

It should not be a surprise then that the principal characteristic of these works is the constant presence of nature, be it through exotic fauna and exuberant vegetation, or in the form of geological layers, soil, strata, creases, anticlines, cuts, rocks, and underground sandstones, which must be studied in depth in order to reach their most valuable secret: oil. This nature is expressed in writings mostly as an obstruction to civilization, a fiend that must be dominated by the protagonists of this modern history: engineers, geologists, and drillers, "Our Boys", "None [of them, writes Ralph Arnold], escaped the ravages of nature in this rugged land" (p. 229).

The author of *Los antecesores* also expresses with great conviction that "the jungle and the marshes were still there, but the men found ways of domesticating them" (p. 209).[4] This task of domesticating the woods and marshes included violent combats with indigenous peoples, especially with the Motilones, the tough Bari ethnic group who, since remote times inhabits the banks of the Catatumbo river on both sides of the frontier between Colombia and Venezuela. They were the ones most affected by the exploratory missions in Sierra de Perijá, as was the jungle, with nature as their main protective shield. In this region, the Motilones tried constantly

---

4 Anonymous: *Los Antecesores: Origen y Consolidación de una Empresa Petrolera*. Departamento de Relaciones Públicas, Lagoven: Caracas 1989.

to repel the missions of the geologists with bows and arrows, which made exploration work very difficult and produced many deaths on both sides. This situation was acknowledged by the young explorers, as a sort of "justifiable desire for revenge on the part of Venezuelans" (Arnold et al., p. 275). He comments, "Can you blame them? We were uninvited guests in their country" (Anonymous, p. 143).

This attitude of the American geologists—mostly young men in their last years of college, or recently graduated from universities like Stanford or Harvard—is frequent in both texts. This reveals, on one hand, their unawareness of the contracts between the oil companies and the government, which guaranteed—with arms, if necessary—the protection of the explorers before any attack of the "savage tribes"; and, on the other hand, as a sort of ecological awareness, it becomes evident that, in many cases, the "yankees" were not the major threat to the original population of the jungle, but rather that the worst enemies were the locals themselves. "This was due to the Venezuelans having rounded up numbers of them and forced them to labor as slaves gathering rubber in the jungle", as happened with the Guaraunos who inhabited the delta of the Orinoco river. Hence, this scenario accounts for the countless problems the geologists encountered when employing indigenous people in their field work, since they hardly had any news of the arrival of strangers and fled to the depths of the jungle to hide. Richard Conkling describes how "after we had employed one or two and given them canned goods and cigarettes as well as paying for their labor, we were overwhelmed by them" (p. 239).

These rather ambivalent views of the narrators evince one of the most interesting aspects of these texts, which is part of a literary production carried out from the very interior of the modernizing, nature-destroying enterprise.

In this sense, the texts offer a very original viewpoint that has always been equivocal in literature. I am referring to the fictionalized chronicle of some processes coupled not only to economic, political or social circumstances, but also to human ones, in a phase of accelerated environmental transformation determined by the oil industry. As a consequence, the chronicles compiled by Arnold, Macready, and Barrington, like those of the author of *Los antecesores*, answer to this concern set forth by Amitav Ghosh when he states, in his attempt to critique the American writers in their impossibility of writing the great oil novel, "there isn't very much they could write about: neither they nor anyone else really knows anything at all about the human experiences that surround the production of oil" (p. 140).[5]

---

5 Ghosh, Amitav: "Petrofiction. The Oil Encounter and the Novel". *Incendiary Circumstances. A Chronicle of the Turmoil of Our Times*. Houghton Mifflin Company: Boston 2005, pp. 138–151.

Taking into consideration only the literary production of the Arab world—and a part of American production, Ghosh errs in a flagrant manner when he ignores Latin America entirely—a territory that, at the beginning of the twentieth century, became the first oil exporter in the world, he concludes, "it's no accident, then, that the genre of 'My Days in the Gulf' has yet to be invented" (p. 140). Quite the contrary, *The First Big Oil Hunt* and *Los antecesores* achieved important innovations in the configuration of the emerging petro-world, furthermore preparing the field for what would later be developed by Latin American oil-encounter fictions. As a result, these two works must undoubtedly be considered as important forerunners of what Graeme MacDonald (2017) states is the subgenre of world literature on oil and incorporated as key pieces of this original bioenergetic genealogy that has begun to be grasped around cultural and literary studies, energy and environmental humanities, and ecocriticism.[6]

Petrofictions: Oil-Encounter Novels in Latin America

At the same time that *Oil!* by Upton Sinclair (1927) was being published in the States, oil literature was also making its appearance in Latin America, only a few years after the region entered unexpectedly into a new epochal cycle determined by the conversion of the old "mene" or "chapopote"—as indigenous people from Venezuela and Mexico called the foul-smelling black oil—into the essential element for modern life. And although these Latin American petrofictions have been barely visible in international scenarios and are fated to remain almost entirely forgotten by local critics, it is unquestionable that they have been one of the most effective channels through which the nature of oil in Latin America has been addressed. The reason for this conclusion is that the literary discourse was able to reconstruct the radical transformations of space and changes in mentalities, experience, and tactics that the new subsoil ecology introduced to vast regions of the continent. It is a new bioenergetic phase determined by the ascent of fossil energy to a global scale (Szeman) that Stephanie LeMenager calls "petroworld" and Mirna Santiago describes as a new "ecology of oil".[7]

---

6 Macdonald, Graeme: "'Monstruos transformer': Petrofiction and world literature". *Journal of Postcolonial Writing* 53(3), 2017, pp. 289–302.
7 Szeman, Imre: "Introduction to Focus: Petrofictions". *American Book Review*, March April 2012, p. 3. LeMenager, Stephanie: *Living in Oil: Petroleum Culture in the American Century*. Oxford University Press: New York 2014. Santiago, Mirna: *The Ecology of Oil: Environment, Labor, and the Mexican Revolution, 1900–1938*. Cambridge University Press: Cambridge 2006.

In these texts, the value conferred in the narrations can be clearly seen as a form of representing reality, and it becomes evident in the desire to organize entirely and coherently an image of life, an objective that can only be obtained in an imaginary form, as was expressed by Hayden White when he studied the role of literature in history.[8] One of the essential characteristics of the first Latin American petrofictions was the disposition to novelize, thereby accepting the responsibility of a chronicle of the times. In addition, this tendency produces a sort of oil realism in the end, cornering some works that, as Venezuelan critic Miguel Angel Campos states, are truer to reality than to literary realism and eventually run aground in the desire to reconstruct history and educate regarding consequences, rather than produce literature (2005, pp. 8–9).[9]

Among the initial works of the Latin American oil-encounter fictions, the following titles are found only in Mexico, and they are listed in chronological order and up to what I have been able to investigate: *Tampico* (1926) by the American Joseph Hergesheimer, *Mapimí 37* (1927), Mauricio Magdaleno's first novel, and *La Hermana impura* (1927) by Jose Manuel Puig Casauranc. In 1928, the extraordinary *Panchito Chapopote* by Xavier Icaza appears, and a year later *Die Weisse Rose* (1929) by B. Traven—translated to Spanish as *La Rosa Blanca* in 1940 and successfully made into a film in 1961. In 1939, *Resaca* by César Garizurieta and *Huasteca* by Gregorio López Fuentes are also published.[10]

---

8 White, Hayden: "The Value of Narrativity in the Representation of Reality". In: Mitchel, W.T.J. (ed.): *On Narrative*. The University of Chicago Press: Chicago 1981, pp. 1–23.
9 Campos, Miguel Angel: "Narrativa del petróleo: evidencias y acuerdos". Introductory study to: Carrera, Gustavo Luis: *La novela del petróleo*. Universidad de los Andes: Mérida, Venezuela. 2005.
10 Hergesheimer, Joseph: *Tampico: a novel*. Alfred A. Knopf: New York 1926; Magdaleno, Mauricio: *Mapimí 37*. Excelsior: Mexico 1927; Puig Casauranc, José Manuel: *La hermana impura*. Editorial Cultura: Mexico 1927; Icaza, Xavier: *Panchito Chapopote (Retablo tropical o relación de un extraordinario sucedido en la heróica Veracruz)*. Editorial Cultura: Mexico 1928; Traven, B.: *Die weisse Rose*. Büchergilde Verlag: Berlin 1929; Garizurieta, César: *Resaca*. Editorial Dialéctica: Mexico 1939; López Fuentes, Gregorio: *Huasteca*. Ediciones Botas: Mexico 1939. For a more detailed revision of the oil novel in Mexico, I refer to the works by Rapp, Helen Louise Rapp: *La novela del petróleo en México*. Universidad Autónoma de México: Mexico 1957; Mario Schneider, Luis: Schneider, Luis Mario: "La literatura del petróleo en México". In: Reyes, Agustín / Tejedo, Lorea San Martín (eds.): *México: a cincuenta años de la expropiación petrolera*. Universidad Nacional Autónoma de México: Mexico 1989, pp. 295–308; and Negrín, Edith: "En los inicios de la novela mexicana del petróleo: La hermana impura de José Manuel Casauranc". In: Jiménez de Báez, Yvette (ed.): *Varia lingüística y literaria. 50*

In other scenarios on the continent, two Colombian authors, Rafael Jaramillo Arango and César Uribe Piedrahita, publish during the decade of the thirties *Barrancabermeja* (1934) and *Mancha de Aceite* (1935), on which I will comment later. In 1936, *Paralelo 53 sur* by the Chilean Juan Marín appears.[11]

The Venezuelan case is exceptional; albeit being one of the oldest petrostates in the world, the oil imaginary barely appears in literary works. Hence, Antonio López Ortega has come to refer to oil as an "absent category" of the literary production (p. 112).[12] This bizarre scarceness of petrofictions also caught the attention of Amitav Ghosh when he observed other oil regions of the world and asked "why, when there is so much to write about, had this encounter proved so imaginatively sterile?" (p. 139). And Campos adds that Ghosh noticed that there was much more "in the process of oil economy that in its imaginary" (2005, p. 8). As a result, one must resort to a set of works that, although of a very irregular quality, nonetheless offers an interesting approach to the nature and scenarios of oil.

Among these works it is necessary to mention, at least for its innovative character, *Lilia: ensayo de novela venezolana* (1909) by Ramón Ayala. Although to be fair, and following a detailed study by Gustavo Luis Carrera (1972), the first oil novel in Venezuela is *El señor Rasvel* (1934) by Miguel Toro Ramírez. To this must be added the abovementioned *Mancha de Aceite* (1935) by Uribe Piedrahita, set in Venezuela; the prized *Mene* (1936) by Ramón Díaz Sánchez, *Guachimanes* (1954) by Gabriel Bracho Montiel, *Los Riberas* (1957) by Mario Briceño-Iragorry, and *Oficina No 1* (1961) by Miguel Otero Silva.[13]

---

años del Centro de Estudios Lingüísticos y Literarios. El Colegio de México: Mexico 1997, pp. 409–426.
11 Jaramillo Arango, Rafael: *Barrancabermeja*. Editorial E.S.B.: Bogotá 1934; Uribe Piedrahita, César: *Mancha de Aceite*. Editorial Renacimiento: Bogotá 1935; Marín, Juan: *Paralelo 53 sur: novela*. Editorial Nacimiento: Santiago de Chile 1936.
12 López Ortega, Antonio: *Discurso del subsuelo*. Oscar Todtmann Editores: Caracas. 2002.
13 Ayala, Ramón: *Lilia (ensayo de novela venezolana)*. Tipografía Americana: Caracas 1909; Carrera, Gustavo Luis: *La novela del petróleo*. Universidad de los Andes: Mérida, Venezuela 2005 [1972]; Toro Ramírez, Miguel: *El señor Rasvel*. Editorial Universal: Caracas 1934; Uribe Piedrahita, César: *Mancha de Aceite*. Editorial Renacimiento: Bogotá 1935; Díaz Sánchez, Ramón: *Mene*. Cuarto Festival del libro venezolano: Caracas 1958 [1936]; Bracho Montiel, Gabriel: *Guachimanes: doce aguafuertes para ilustrar la novela del petróleo*. Seremos: Santiago de Chile 1954; Briceño-Iragorry, Mario: *Los Riberas: historias de Venezuela*. Editorial Independencia: Madrid 1957; Otero Silva, Miguel: *Oficina N° 1*. Editorial Lozada: Buenos Aires 1961. For a

In this list that only covers the first part of the last century, we should not leave out the countless oil-encounter fictions written by non-Latin American authors, but which take place in Latin America. From my point of view, these works, beyond their plentiful or scarce literary qualities, are of great importance to understand in all its complexity a phenomenon that from the region of literary creation is intrinsically linked to the elaboration of outstandingly global geo-ecological imaginaries.

I have already mentioned the works by American Joseph Hergesheimer and German B. Traven; the latter, a classic of the genre in Mexico where he is considered a countryman. And although I don't wish to make this list much longer, I would like to include here the work of Karl May's *Der Ölprinz*, published in 1888. May is perhaps the most popular author of German literature for his books on journeys and adventures; his book *Der Ölprinz* is set in Mexico, and in it can be perceived some of the elements that will define the oil novel. In Mexico also, *Back River* (1934) by American Carleton Beals, *Chapopote: Der Götze Mexikos* [*Chapopote: The God of Mexico*] (1936) by Victor Pfeiffer and Emil Marboth are developed; as well as *Chapopote* (1946) and *Tepetate* (1947) by the Austrian geologist and writer Hans Adrian. Finally, in Venezuela, there is a short novel by the German Hans Hoffmann (1958), *Ölsucher am Rio Catatumbo* [*Oil Hunters on the Catatumbo River*], and the very important *El salario del miedo* [*The Wages of Fear*] by the French writer, Georges Arnaud (1950), which turned the genre upside down, and on which I will comment later.[14]

This first petroliterature concentrates mostly on observing how the ascent and consolidation of the oil ecology translated in these regions into a violent structural reorientation that summarizes the transition from an agrarian, semi feudal economy, to an economy determined by the modern oil enclave. This process implied not only the acceleration of the economy subordinated to the metropolis, but it also generated, as a consequence, important changes in property and land use—"the commodification of nature" (Santiago 2006, pp. 61–100)—and as

---

study of the oil novel in Venezuela, see Carrera, Gustavo Luis (1972) and Campos, Miguel Ángel (1994, 2005).

14 May, Karl: *Der Ölprinz*. Karl-May-Verlag: Bamberg 2001 [1888]; Beals, Carleton: *Back River*. J.P. Lippincott Company: Philadelphia 1934; Pfeiffer, Victor / Marboth / Emil: *Chapopote: Der Götze Mexicos. Das Bergland Buch*: Salzburg 1936; Adrian, Hans: *Chapopote: eine Erzählung um Mexikos Erdöl*. E. Reinhardt: Basel 1946; Adrian, Hans: *Tepetate: Auf Ölsuche im mexikanischen Busch*. E. Reinhardt: Basel E. Reinhardt 1947; Hoffmann, Hans: *Ölsucher am Rio Catatumbo*. J.G. Oncken Verlag: Kassel 1958; Arnaud, Georges: *Le salaire de la peur*. René Julliard: Paris 1950.

Miguel Tinker Salas (2009) evaluated, a geographic transformation of the landscape, which included the configuration of new population groups due to the massive migration to the enclaves, new forms of labor, as well as new patterns in social divisions.[15] Generally, the process refers to the incorporation of the continent to original life-style and cultural habits molded by what Carlos Monsiváis describes as a crushing Americanization.

If in the accounts we saw above, the key lies in the look of the other, the vision of an exotic culture through the eyes of a foreigner, in the oil novels—all of them written from a local standpoint—what prevails is the hallucinated vision of the radical life transformations that occur after the arrival of the foreigners. With them came the conversion of the nature of the subsoil into an unfathomable wealth never before known.

We can see this clearly in the work of Xavier Icaza, *Panchito Chapopote: Retablo tropical o relación de un extraordinario sucedido en la heróica Veracruz* [*Panchito Chapopote: Tropical Tableau or An Extraordinary Event in Heroic Veracruz*]; this rather obscure writing is nonetheless considered by some as "one of the most extraordinary phenomena in the study of Mexican literature" (Brushwood, p. 161).[16] The work, identified with expressive procedures of the Mexican strident movement, is one of the starting points of the Latin American oil narrative tradition since in it are found almost all the elements we will see throughout an entire century of literary production dedicated to the subject: the testimonial character of the works, the topics and approaches (anti-imperialist nationalism, economic transformations, unstoppable modernization, corruption, discrimination, peasant emigration, the company's superpower, degradation of nature), the backgrounds (jungle, farm, oil wells, oil fields), and the archetypal characters (geologist, engineer, driller, company, peasant, workers, prostitutes).

Panchito Chapopote takes place at the beginning of the twentieth century in the state of Veracruz during the last period of Porfirio Díaz's political control, and the beginning of the Mexican revolution. Panchito, the main character, is a peasant heir to some lands unfit for agriculture since from the bottom of that land spurts a "chapopotera" that contaminates the water and "burns the trees". But his life takes an unexpected turn after reports are made public stating that the oil from his orchard signals the existence of a hidden wealth in the intricacies of the ground.

---

15 Tinker Salas, Miguel: *The Enduring Legacy: Oil, Culture and Society in Venezuela.* Duke University Press: Durham 2009.
16 Brushwood, John S.: "Las bases del vanguardismo en Xavier Icaza". *Texto Crítico* 8 (24-25), 1982, pp. 161-170.

> Panchito didn't have any possessions. Better still, as if he had none. His father had left him some lands, but you could say they had been damned by God. [...] They didn't produce anything. The seeds were fruitless in the tarry lands. The water from the spring gushing in them was oily. It looked damned. One could say that its simple contact burned the plants. (pp. 17–18)[17]

With this argument that also inaugurates the vision of oil as cursed wealth, "the devil's breeding ground" by Mexican poet Ramón López Velarde, or "the Devil's dung" in the sociological version by the Venezuelan Juan Pablo Pérez Alfonzo, the work focuses on describing the socioeconomic transformations and geopolitical conflicts generated by an abrupt epochal change described as aggressively disruptive. A change expressed in the novel as a testimonial that, as we have mentioned, will constitute one of the essential features of the first oil literature: its obsession with recording a novelty, and at the same time, creating an awareness before what is considered plundering.

In this approach, however, no ecological awareness is present, although an eco-critical thought is expressed in the political questioning of the commodification process of nature and the consequences this has on native populations. This opinion is of great value for contemporary ecological approaches, since as Cheryl Lousley points out, the challenges of eco-criticism in the field of literature are not limited to the representation of nature but go beyond to the politicization of the environment to ask how to make socioecological conflicts socially visible as a political issue (p. 156).[18] Hence, and as we will also see in the following works, this oil literature engages not only with the tradition of the imperialism critique, the formulation of late nationalisms, and the later tenets of the theory of dependency, and postcolonial perspective frequently used when approaching the oil problematic, but it also connects with the ecocritical thought that has served to open new ways to studies produced from and about Latin America.

Another almost unknown work outside the academic cycle is *Mancha de aceite* [*Oil Stain*] by the Colombian César Uribe Piedrahita. This work, together with *Panchito Chapopote,* signals the foundational moment of the oil literary canon in Latin America. For this reason, Gustavo Luis Carrera considers it "the first novel [...] founded entirely on the oil issue, its ambience, characters, problems and perspectives" (2005, p. 50).

---

17 My own translation, as well as all the following quotes.
18 Lousley, Cheryl: "Ecocriticism and the Politics of Representation". In: Garrard, Greg (ed.): *The Oxford Handbook of Ecocriticism.* Oxford University Press: New York 2014, pp. 155–171.

As in Icaza's novel, this is also a writing of a great documentary character, interested in carrying out a record of the strategies put forth by the transnational oil companies in their endeavor to colonize nature on the Latin American horizon. However, unlike Icaza, Uribe Piedrahita concentrates on fictionalizing human drama: the precarious work conditions where the system of oil extraction in the forest was put in motion, just as was expressed by Ralph Arnold in his chronicles; likewise, how this translated into a radical change in the life conditions of the peasants as they became incorporated in it, as well as the indigenous peoples who were displaced, and according to the account, exterminated by systematic bombings of the region.[19]

Thus, in a dramatic manner that contrasts with the adventure tales of the American geologists, the description is given of the configuration of a sort of human destruction machine in which converge the interests of the dictatorship of Juan Vicente Gómez with the business interests of the emerging global capitalism:

> In the pools, imprisoned by the rods and sleepers of the giant skeleton, Gustavo saw shoulders, hips and heads. Someone moaning in the sludge and further on someone else asking "the Virgin" to pull him out soon and let him live [...]
> Night fell and in the light of the reflectors and lanterns the injured were placed in series and the fragments of men further away. Formless masses, destroyed faces, viscera covered in sludge, torn bellies, meaty ruins. Ailing and crushed meat! Six, seven, nine... Some still missing and the arroyo increased. (Uribe Piedrahita, p. 10)

Thus abound descriptions of poverty transplanted to the ancient oil fields, surviving in the swampy jungle: "(in) this land there are no roads, no fields...no women, no anything, right, doctor?" (Uribe Piedrahita, p. 6). But more crucially, the tragedy of the countless deaths produced by the frequent accidents of an industry in an experimental phase; likewise, malaria, yellow fever, all of which manifest the unhealthiness of the emerging oil ecology—"no one goes out because everyone is fevered" (Uribe Piedrahita, p. 52).

One of the most striking writing resources of *Mancha de Aceite* is the description of nature as a subject affected by a tenebrous illness. Using metaphors that project human pathologies on nature, Uribe Piedrahita—a doctor to-be—places at the service of literature a representation of oil that is genuinely anthropocentric,

---

19 Regarding the drama of the indigenous population in Western Venezuela during the process of displacement on the part of the oil companies, see Alarcón Puentes, Johnny: "Indígenas y empresa petrolera a principios del siglo XX. Origen de una disputa". *Boletín Antropológico*, 63, 2005, pp. 31–55.

humanizing the environment through the description of the clinical etiology and manifestations of nature transformed into a sick patient.

> Downwards ran a sulfurous rivulet belching foul, hot gases. [...] Soundlessly, or perhaps the earth moaned, its lament lost in the wheezing of the motors exploding and the shouts of the foremen. Noiselessly, slowly, the hills leaning on the tilted plane of the cracked rock, caved in due to its height, and slid down becoming crumpled at the base where the large feet of the tower squeezed the breast of the slope. (pp. 9–10)

With this orientation, the discovery of an oilfield is described as lancing a purulent abscess. It is not an oil well ejecting bitumen from the subsoil; rather, it is an inflamed ulcer oozing a bilious liquid:

> The earth bellowed and quavered to its armpits. The soil shook and with loud pushes, and rasps, it vomited all along the pipes, against the tower, and up to the sky, a black belch that dragged with it the martinet, the cables, sleepers, beams and twisted skeletons of the giant tower, and bristled its ironworks, wires and broken sheets on the side. It looked as if the asphaltic oil gobs reached the sun itself. (p. 29)

This imaginary that describes the emergence of radically new socioecological landscapes will be developed in *Mene* by Ramón Díaz Sánchez, and some years later, in *Oficina N° 1* by Miguel Otero Silva. In *Mene*, Díaz Sánchez relies on his own experience in the city of Cabimas to expound on the transformation to the new modern poverty that is emerging around the oilfield, dragging with it huge migratory groups attracted by the new myth of progress:

> The crews grew constantly, organized under a steely, machine-like discipline. Now it wasn't only blonds and indigenous peoples on the gnawed land. Each morning new boats filled with strange men with strange languages and strange colors would arrive. Babel personified its myth on this fiery piece of land. They all came with the same fever, the same longings. Dark towns—Cabimas, Lagunillas—became part of the world's flurry. Footpaths became streets, shacks became homes: precarious dwellings, built with boxes of the machinery and covered with zinc sheets. The madness of a dream taken from illusory frontiers. (p. 30)

It was thus that the jungle and the swamps' silence suddenly had to deal with the noise of modern machines: steam boats, airplanes, refrigerators, electric stoves—crudely, the paradigm that rules reality is another one: everything can be electric—and the uproar of entire villages and towns transformed into oilfields:

> Each drill has a drill that suctions [...] each drill has a motor that palpitates [...] each motor has a boiler that regurgitates like a monstrous open artery. Beside all this in the premises of "*El Hijo de la Noche*" there were a thousand mouths screaming and laughing, two thousand plants stomping, a derelict orchestra screeching desperately, destroying a *paso-doble*, and a thousand fists knocking on doors, tabletops and iron chairs. From the street came the roar of the automobiles, and the hurt cry of the gramophones. (p. 54)

The violent and traumatic transition to the modern world is also Miguel Otero Silva's subject in *Oficina No 1*. Beginning with the birth of an oil town, this novel narrates the origin of the peasants' exodus in search of progress, as well as the foundation of towns that emerged from the prosperity of a time that makes them perish at the same speed it saw them rise, as a metaphor of the decadence of the agricultural landowning system and the emergence of a different model whose determining factor is oil.

However, the singularity of Otero Silva's novel does not lie in pointing out structural changes, but rather the modifications that are taking place in the field of culture, especially day-to-day culture. As Venezuelan writer Juan Liscano observed, without being prepared for it, the entire population entered into a vertiginous metamorphosis that encompassed physical space, but also social behavior, work, exchange media, prices and value of things, habits, entertainment, and even language and speech (p. 17).[20] These changes that clearly show the transition to a totally different culture, to a "petroculture", as indicated by Wilson, Szeman, and Carlson (p. 14), must be observed as part of the incorporation to Western civilization by Latin American culture, "a violent and merciless place that represents the [...] seed of a new future" (Franco, p. 268).[21]

Closing: Global Imaginaries of Latin American Oil Nature

As can be appreciated, from a regional perspective, the corpus of the Latin American petrofictions is wider and more interesting than what is usually believed, and it has not stopped growing in our day with works from noteworthy authors like Mexican Carlos Fuentes, Venezuelan José Balza, and Colombian Laura Restrepo, the only woman on the list. It is even broader if we incorporate the stories and what has been done in theater, poetry, and film, which is not as scarce as believed. As a result, in view of the systematic and structural connections, recurrent motives, topics, scenarios, and archetypal characters, it would not be outlandish to propose the existence of an authentic aesthetics of oil in Latin America, an aesthetics with profound connections to what LeMenager studies in the popular culture and literature in the States (pp. 67–101); this is the same which Graeme Macdonald observes in his revision of the novels *Cities of Salt* by Abdelrahman Munif and *Greenvoe* by George MacKay Brown. Despite their disparities, these writings show how "the unifying resource

---

20 Liscano, Juan: "La industria del petróleo como factor de transculturización". *Geosur* 27, 1981, pp. 12–25.
21 Franco, Jean: 1971. *La cultura moderna en América Latina*. Joaquín Mortiz: Mexico 1971.

system of petro-capitalism forges fundamental cultural affinities and world-ecological connections that cut across and ultimately transcend such disparities" (Macdonald, p. 300). Hence, if in following Macdonald, we consider in all of them recognizable forms in the literary production, and more widely, of oil culture, that from the periphery complete the account of an eminently global story, it would not be inconceivable to consider these Latin American fictions—written at different times, in diverse contexts and with practically no connection between the authors—as part of a subgenre of world-literature: a world-oil-literature.

This Latin American oil imaginary produced by literature remained confined to the continent for almost half a century, or limited to specialists with a very limited diffusion, even among the very same Latin American countries. It was only when the novel *Le salaire de la peur* [*The Wages of Fear*] by the French author Georges Arnaud appeared in 1950, but especially with the huge success achieved by its adaptation to film carried out by Henri-Georges Clouzot, that it reached international visibility.

Le salaire de la peur is the story of a group of lowlife men expelled from Europe during the Second World War who remain trapped in a miserable oil town in Latin America, waiting for a stroke of luck that would allow them to change their fate. As the premise for the events that unfold, the oil company that controls the town's economy offers them a chance to escape from their boring nightmare in exchange for driving two trucks loaded with nitroglycerine, destined to suffocate the fire at an oil well located some 400 kilometers from there. What follows is an extremely valuable representation of the "modern tragedy" that drives the oil exploitation; it is a bioenergetic drama in which Clouzot reflects, with great sensibility, the stories of poverty and destruction of nature that Arnaud and other Latin American authors were able to put forward in literature.

The huge success of the film served, then, to project the oil imaginaries on Latin American soil in the worldwide art and entertainment circuit.[22] It was thus that from the association of the story in which Yves Montand and Charles Vanel starred and, with the chronicles and fictions produced regarding Latin America, an imaginary space emerged that was hugely profitable for the cultural industry of its time. As of this moment, the industry initiated the implementation of a whole literary and cinematographic production whose objectives were not focused on the problems or records of the socioecological complexity of the oil nature, but rather on benefitting commercially from the cultural novelty of the phenomenon.

---

22 The film *Le salaire de la peur* received in 1953 the Palme D'or at Cannes Film Festival, and the Golden Bear at the Film Festival in Berlin; in 1954 the award from the Film Critics Union of France, and in 1955, England's BAFTA award for best picture.

At the same time and in an involutionary manner, this entrepreneurship helped to propagate globally the discourses and imaginaries that for decades fed what I have called "the myth of the modern oil nation" in Latin America.

The production of these oil-oriented accounts set on the continent thus became present in countless pocket-book samples of literature, as well as in B film series. These productions, in which the nature of oil was totally devoid of content, simply became the backdrop for writing all kinds of works with intentions of marginal authorship and no political, much less ecological, awareness. As for the oil companies, the only interest was to exploit the nature of oil as a scenario and not to understand the complex energy-ecological problems.

Beginning with these writings, the expansion of the notion of a Latin America associated with the oil imaginary runs unstoppably in countless texts that helped to build an account of the nature and landscapes of oil. In the same manner, the oil enclave account was built as a limited and exceptional space, isolated from the complex, global economic, and political reality.

As a result, turning our gaze towards the initial works of oil encounter fictions facilitates the approach to a valuable fictional record of the energetic transit that goes from coal ecology to oil ecology; or from the steam engine to the piston motor. At the same time, it offers clues of enormous value for evaluating, through the eyes of literary and cultural history and the more recent ecological debates, the beginning of a cycle of acceleration of the large-scale exploitation of nature: the so-called Anthropocene.

This literature produced during the first half of the twentieth century tells us precisely of the original entanglements present at the center of the new phase of globalization determined by the boom and expansion of fossil energy; it comprises a set of works of unquestionable worth that opens new paths to the imagination, but also to the different branches of humanities and social sciences for rethinking the bioenergetic history of the planet and for reconsidering how the process of accelerated expansion, boosted by subsoil energy, has led to a sort of suicidal transit towards our own destruction.

Bibliography

Adrian, Hans: *Chapopote: eine Erzählung um Mexikos Erdöl*. E. Reinhardt: Basel 1946.

Adrian, Hans: *Tepetate: Auf Ölsuche im mexikanischen Busch*. E. Reinhardt: Basel 1947.

Alarcón Puentes, Johnny: "Indígenas y empresa petrolera a principios del siglo XX. Origen de una disputa". *Boletín Antropológico*, 63, 2005, pp. 31–55.

Anonymous: *Los Antecesores: Origen y Consolidación de una Empresa Petrolera*. Departamento de Relaciones Públicas, Lagoven: Caracas 1989.

Araujo, Orlando: *Narrativa venezolana contemporánea*. Editorial Tiempo Nuevo: Caracas 1972.

Arnaud, Georges: *Le salaire de la peur*. René Julliard: Paris 1950.

Arnold, Ralph / Macready, George / Barrington, Thomas: *The First Big Oil Hunt. Venezuela 1911-1916*. Vantage Press: New York 1960.

Arnold, Ralph / Macready, George / Thomas Barrington: *Venezuela petrolera: Primeros pasos. 1911-1916*. Fundación Trilobita: Caracas 2008.

Ayala, Ramón: *Lilia (ensayo de novela venezolana)*. Tipografía Americana: Caracas 1909.

Beals, Carleton: *Back River*. J.P. Lippincott Company: Philadelphia 1934.

Bracho Montiel, Gabriel: *Guachimanes: doce aguafuertes para ilustrar la novela del petróleo*. Seremos: Santiago de Chile 1954.

Briceño-Iragorry, Mario: *Los Riberas: historias de Venezuela*. Editorial Independencia: Madrid 1957.

Brushwood, John S.: "Las bases del vanguardismo en Xavier Icaza". *Texto Crítico* 8 (24-25), 1982, pp. 161-170.

Campos, Miguel Angel: "Narrativa del petróleo: evidencias y acuerdos". Introductory study to: Carrera, Gustavo Luis: *La novela del petróleo*. Universidad de los Andes: Mérida, Venezuela 2005.

Campos, Miguel Ángel: *Las novedades del petróleo*. Fundarte: Caracas 1994.

Carrera, Gustavo Luis: *La novela del petróleo*. Universidad de los Andes: Mérida, Venezuela 2005 [1972].

Clouzot, Georges: *Le salaire de la peur* (Film). Compagnie Industrielle et Commerciale Cinématographique, Filmsonor, Vera Films: France/Italy 1953.

Díaz Sánchez, Ramón: *Mene*. Cuarto Festival del libro venezolano: Caracas 1958 [1936].

Franco, Jean: *La cultura moderna en América Latina*. Joaquín Mortiz: México 1971.

Garizurieta, César: *Resaca*. Editorial Dialéctica: México 1939.

Glotfelty, Cheryll / Fromm, Harold (eds.): *The Ecocriticism Reader: Landmarks in Literary Ecology*. University of Georgia Press: Athens, Georgia 1996.

Ghosh, Amitav: "Petrofiction. The Oil Encounter and the Novel". In: *Incendiary Circumstances. A Chronicle of the Turmoil of Our Times*. Houghton Mifflin Company: Boston 2005 [1992], pp. 138-151.

Heffes, Gisela: "Para una ecocrítica latinoamericana: entre la postulación de un ecocentrismo crítico y la crítica a un antropocentrismo hegemónico". *Revista de Crítica Literaria Latinoamericana* 79, 2014, pp. 3-6.

Hergesheimer, Joseph: *Tampico: a novel*. Alfred A. Knopf: New York 1926.

Hoffmann, Hans: *Ölsucher am Rio Catatumbo*. J.G. Oncken Verlag: Kassel 1958.

Icaza, Xavier: *Panchito Chapopote (Retablo tropical o relación de un extraordinario sucedido en la heróica Veracruz)*. Editorial Cultura: México 1928.

Jaramillo Arango, Rafael: *Barrancabermeja*. Editorial E.S.B.: Bogotá 1934.

LeMenager, Stephanie: *Living in Oil: Petroleum Culture in the American Century*. Oxford University Press: New York 2014.

Liscano, Juan: "La industria del petróleo como factor de transculturización". *Geosur* 27, 1981, pp. 12–25.

López Fuentes, Gregorio: *Huasteca*. Ediciones Botas: Mexico 1939.

López Ortega, Antonio: *Discurso del subsuelo*. Oscar Todtmann Editores: Caracas 2002.

López Velarde, Ramón: *La suave patria y otros poemas*. Fondo de Cultura Económica: México 1983.

Lousley, Cheryl: "Ecocriticism and the Politics of Representation". In: Garrard, Greg (ed.): *The Oxford Handbook of Ecocriticism*. Oxford University Press: New York 2014, pp. 155–171.

Macdonald, Graeme: ""Monstruos transformer": Petrofiction and world literature". *Journal of Postcolonial Writing* 53(3), 2017, pp. 289–302.

Magdaleno, Mauricio: *Mapimí 37*. Excelsior: México 1927.

Mandrillo, Cósimo: "Intimidad y nostalgia en *De un pueblo y sus visiones* de J.M. Villarroel París". Introduction to Villarroel París, José Miguel: *De un pueblo y sus visiones*. Fondo Editorial del Caribe: Barcelona, Anzoátegui 2009, pp. 7–11.

Marín, Juan: *Paralelo 53 sur: novela*. Editorial Nacimiento: Santiago de Chile 1936.

May, Karl: *Der Ölprinz*. Karl-May-Verlag: Bamberg 2001 [1888].

Monsiváis, Carlos: *Aires de familia: cultura y sociedad en América Latina*. Anagrama: Barcelona 2000.

Negrín, Edith: "En los inicios de la novela mexicana del petróleo: La hermana impura de José Manuel Casauranc". In: Jiménez de Báez, Yvette (ed.): *Varia lingüística y literaria. 50 años del Centro de Estudios Lingüísticos y Literarios*. El Colegio de México: México 1997, pp. 409–426.

Ordaz, Ramón: *Piedra de aceite: Recepción del tema petrolero en la poesía venezolana*. Fondo Editorial del Caribe: Barcelona, Anzoategui 2012.

Otero Silva, Miguel: *Oficina N° 1*. Editorial Losada: Buenos Aires 1961.

Pérez Alfonzo, Juan Pablo: *Hundiéndonos en el excremento del diablo*. Editorial Lisbona: Caracas 1976.

Pfeiffer, Victor / Emil Marboth: *Chapopote: Der Götze Mexikos*. Das Bergland Buch: Salzburg 1936.

Puig Casauranc, José Manuel: *La hermana impura*. Editorial Cultura: México 1927.

Rapp, Helen Louise: *La novela del petróleo en México*. (master thesis) Universidad Autónoma de México: México 1957.

Rivera, Hesnor: *Persistencia del desvelo*. Monte Avila Editores: Caracas 1976.

Santiago, Mirna: *The Ecology of Oil: Environment, Labor, and the Mexican Revolution, 1900–1938*. Cambridge University Press: Cambridge 2006.

Schneider, Luis Mario: "La literatura del petróleo en México". In: Reyes, Agustín / Tejedo, Lorea San Martín (eds.): *México: a cincuenta años de la expropiación petrolera*. Universidad Nacional Autónoma de México: Mexico 1989, pp. 295–308.

Schneider, Luis Mario: *La novela mexicana entre el petróleo, la homosexualidad y la política*. Nueva Imagen: México 1997.

Silva-Ferrer, Manuel: "Venezuela: the Modern Oil Nation. Globalización, Estado, cultura y comunicación en torno al enclave petrolero". *Tiempo y Espacio* 33(63), 2015, pp. 35–53.

Sinclair, Upton: *Oil!*. Author: Long Beach, California 1927.

Szeman, Imre: "Introduction to Focus: Petrofictions". *American Book Review* March-April, 2012, p. 3.

Tinker Salas, Miguel: *The Enduring Legacy: Oil, Culture and Society in Venezuela*. Duke University Press: Durham 2009.

Toro Ramírez, Miguel: *El señor Rasvel*. Editorial Universal: Caracas 1934.

Traven, B.: *Die weisse Rose*. Büchergilde Verlag: Berlin 1929.

Uribe Piedrahita, César: *Mancha de Aceite*. Editorial Renacimiento: Bogotá 1935.

White, Hayden: "The Value of Narrativity in the Representation of Reality". In: Mitchel, W.T.J. (ed.): *On Narrative*. The University of Chicago Press: Chicago 1981, pp. 1–23.

Wilson, Sheena; Adam, Carlson, and Imre Szeman (eds.): *Petrocultures: Oil, Politic, Culture*. McGill-Queen's University Press: Montreal 2017.

Contributors

Natalia Álvarez Méndez is Tenured Professor of Comparative Literature and Literary Theory at the Universidad de León. She has published *Espacios narrativos* (2002) y *Palabras desencadenadas. Aproximación a la teoría literaria postcolonial y a la escritura hispano-negroafricana* (2010) and a number of articles and book chapters. As the director of the research group Grupo de Estudios Multitextuales de lo Insólito y Perspectivas de Género (EMIPG) of the Universidad de León, she has coordinated two editions on this field of research, *Espejismos de la realidad. Percepciones de lo insólito en la literatura española (siglos XIX-XXI)* (2015) and *Territorios de la imaginación. Poéticas ficcionales de lo insólito en España y México* (2016).

Arturo Arias is Professor and distinguished John D. and Catherine T. MacArthur Chair at the University of California, Merced. After spending most of the 1980s writing, teaching, and conducting research in Mexico, Arias became a professor at San Francisco State University, where he earned acclaim for his fiction writing. Later, at the University of Texas at Austin, he gained further recognition for his research and scholarly writing, and in 2013, he was named the Tomás Rivera Professor of Spanish Language and Literature. Most recently, Arias has focused his research on the indigenous people of Guatemala and Mexico and the ways in which they are rising to prominence and gaining political and academic influence in Central American society. Arturo Arias has also written seven novels, with an eighth forthcoming. He co-wrote the screenplay for the film "El Norte", and in 2008 won Guatemala's highest literary prize, the Miguel Angel Asturias National Literary Award for Lifetime Literary Production.

Laura Barbas-Rhoden is Associate Professor of Spanish at Wofford College. Her professional work centers on Ecocriticism and the Environmental Humanities; Interdisciplinary, Integrative and Collaborative Practices in Higher Education; and twentieth and twenty-first century Latin American Literature.

Scott DeVries is Associate Professor at Manchester University. His continuing research encompasses both nineteenth and twentieth century Latin American literature from the perspective of ecological criticism, animal studies, and energy. His first book, *A History of Ecology and Environmentalism in Spanish American Literature* with Bucknell University Press was released in 2013 and his most recent, *Creature Discomfort: Fauna-criticism, Ethics, and the Representation of Animals in Spanish American Fiction and Poetry* was published by Brill in

2016. He also has publications in several other venues, including articles on Mexican film in *Christian Scholars Review* and on Jorge Luis Borges in the *Journal of Philosophy and Literature*.

Gisela Heffes is Associate Professor of Latin American Literature and Culture at Rice University. She is the author of two monographs; *Las ciudades imaginarias en la literatura latinoamericana* (2008) is a study of the literary representations of non-existent urban spaces and their significance in the wider political and cultural framework of Latin America, and *Políticas de la destrucción / Poéticas de la preservación. Apuntes para una lectura (eco)crítica del medio ambiente en América latina* (2013) examines narratives from the mid-twentieth century to the present that are related to the environment in Latin America and analyzes how these texts refer to both the conservation and destruction of nature. She has also edited anthologies and essay collections and served as the guest editor for the special issue of *Revista de Crítica Literaria Latinoamericana* on "Ecocrítica" (2014). Dr. Heffes is now serving as Co-Diversity Officer at the Association for the Study of Literature and Environment (ASLE). In addition to her academic work, she is an active fiction writer, having published several novels and short stories.

Jorge Marcone is Associate Professor at Rutgers University. His research interests lie in Environmental Humanities in Latin America and Spain and the far-reaching implications of local or planetary environmental crises and conflicts for ontology, epistemology, ethics, hermeneutics, and aesthetics; in the impact of popular and indigenous environmentalisms in current ecological thinking; and in the literature, film and arts of Amazonia and its prominent role in the planet's ecology and ecological imagination. Prof. Marcone has also published on ecology and the Spanish American Regional Novel, Mexican Literature, Chicana Literature, Pablo Neruda, and José Emilio Pacheco among other authors.

José Manuel Marrero Henríquez is Tenured Professor of Comparative Literature and Literary Theory at the Universidad de Las Palmas de Gran Canaria. He has published extensively on landscape and animal representation in literature and on a variety of topics and authors of the Spanish and Latin American literary traditions. Editor of *Transatlantic Landscapes: Environmental Awareness, Literature and the Arts* (UAH 2016), and co-editor of *Humanidades Ambientales. Pensamiento, Arte y Relatos para el Siglo de la Gran Prueba* (Catarata 2018), his most recent endeavor of a *Poetics of Breathing*, a general ecocritical theory, also resonates within his poems in *Landscapes with Donkey / Paisajes con burro* (Green Writers Press 2018).

Contributors 249

Pamela Phillips is Associate Professor at Universidad de Puerto Rico, Río Piedras. She has published extensively on eighteenth-century Spanish literature and travel literature from and about Spain in the eighteenth, nineteenth, and twentieth centuries.

Beatriz Rivera-Barnes is Associate Professor at Penn State University. She is the co-author (with Dr. Jerry Hoeg) of *Reading and Writing the Latin American Landscape* (Palgrave Macmillan 2009), an ecocritical approach to American literatures. Dr. Rivera-Barnes is also the author of a collection of short stories and four novels published by Arte Publico Press, University of Houston. In 2007, she was awarded the International Latino Book Award for her novel, *Do Not Pass Go*, and was finalist for the Paterson Fiction Prize. Both her creative and scholarly works have been widely anthologized.

Manuel Silva-Ferrer is a researcher in the Instituto de Estudios Latinoamericanos de la Freie Universität Berlin, where he studies the relations between oil, society and culture in Latin America. His academic publications and journalistic collaborations also explore the fields of politics, culture, and media in Latin America. Silva-Ferrer is Associate Researcher of the Instituto de Investigaciones de la Comunicación (ININCO) of the Universidad Central de Venezuela and a member of the committee of the journal *Anuario ININCO*. His career began in television and film production and as a journalist for Radio Caracas Radio. He became a member of the Cinemateca Nacional de Venezuela in 1995 and a member of the Department of Cine and Medios de Comunicación of the Ministerio de la Cultura in 2005. He was also a member of the Centro Nacional Autónomo de Cinematografía de Venezuela and a representative before the Conferencia de Autoridades Cinematográficas de Iberoamérica (CACI). Silva-Ferrer was also the director of the prestigious journal of Latin America photography *ExtraCámara*. His most recent book *El cuerpo dócil de la cultura* (Iberoamericana Vervuert 2014) studies the cultural changes in Venezuela during the "revolución bolivariana".

Studien zu Literatur, Kultur und Umwelt
Studies in Literature, Culture, and the Environment

Herausgegeben von / Edited by
Hannes Bergthaller, Gabriele Dürbeck, Robert Emmett, Serenella Iovino, Ulrike Plath

Bd. / Vol. 1   Daniel A. Finch-Race/Stephanie Posthumus (eds): French Ecocriticism. From the Early Modern Period to the Twenty-First Century. 2017.

Bd. / Vol. 2   Alessandro Macilenti: Characterising the Anthropocene. Ecological Degradation in Italian Twenty-First Century Literary Writing. 2017.

Bd. / Vol. 3   Gabriele Dürbeck/Christine Kanz/Ralf Zschachlitz (Hrsg.): Ökologischer Wandel in der deutschsprachigen Literatur des 20. und 21. Jahrhunderts – neue Ansätze und Perspektiven. 2018.

Bd. / Vol. 4   Yi-Peng Lai: EcoUlysses. Nature, Nation, Consumption. 2018.

Bd. / Vol. 5   Gabriele Dürbeck/Jonas Nesselhauf (Hrsg.): Repräsentationsweisen des Anthropozän in Literatur und Medien / Representations of the Anthropocene in Literature and Media. 2019.

Bd. / Vol. 6   José Manuel Marrero Henríquez (ed.): Hispanic Ecocriticism. 2019.

www.peterlang.com

www.ingramcontent.com/pod-product-compliance
Ingram Content Group UK Ltd.
Pitfield, Milton Keynes, MK11 3LW, UK
UKHW041923210426
5322IPUK00002B/32